Kenneth J. Hsü

Physical Principles of Sedimentology

A Readable Textbook for Beginners and Experts

With 64 Figures

Springer-Verlag Berlin Heidelberg New York
London Paris Tokyo Hong Kong

Professor Dr. KENNETH J. HSÜ
Geologisches Institut
Eidgenössische Technische Hochschule
ETH Zentrum
CH-8092 Zürich

ISBN 3-540-51268-3 Springer-Verlag Berlin Heidelberg New York
ISBN 0-387-51268-3 Springer-Verlag New York Berlin Heidelberg

The use of registered names, trademarks, etc. in this publication does not imply, even in the absence of a specific statement, that such names are exempt from the relevant protective laws and regulations and therefore free for general use.

Printing and bookbinding: Offsetdruckerei Beltz, Hemsbach

2132/3145-543210 – Printed on acid-free paper

Preface

This is a textbook for geology undergraduates taking their first course in sedimentology, for graduate students writing a term paper on sedimentology or preparing for their qualifying examinations, and for instructors, who deem it necessary to infuse a more physical-science approach to the teaching of geology. I also hope that some physics students might find the book readable and comprehensible, and that some of them might be inspired to start a career in the physics of geology.

This textbook is a revision of my lecture notes for my course *Principles of Sedimentology* at the Swiss Federal Institute of Technology. It is a twounit semester course given to second- or third-year undergraduate students, who have acquired a basic knowledge in physics, chemistry, mathematics, and geology. The purpose of teaching this course is to bridge the gap between what has been learned in middle school and in the first year of university to what shall be learned in geology during later years. I intend especially to impart the impression to my students that the study of geologic processes is applied physics, applied chemistry, and applied mathematics. The content of the book is somewhat more extensive than what I have taught, and could be used as a textbook for a three-unit semester course, or even for a two-semester course, if a lecturer chooses to do so. We teach a second course in sedimentology, for which the students are recommended to consult textbooks on depositional environments and facies models. This textbook is not intended to replace, but to supplement those.

James Hutton in the late 18th century and Charles Lyell in the middle of the last century established the natural-history approach to study geology, and the success of the method is witnessed by the progress of the science over the last 2 centuries. The logic is inductive reasoning: Noting that quartz sands are terrigenous detritus derived from deeply weathered terranes, quartz sandstones or orthoquartzites are given paleoclimatic significance. Observing that marine organisms lived, died and are buried in marine sediments, fossil ecology becomes a key to interpreting sedimentary environments. There needs to be no reference to Newtonian mechanics or to Gibbsian thermodynamics, because physical laws cannot be invoked to prove or falsify the interpretation.

The method is necessary, reliable, but, in some cases, insufficient.

Orthoquartzites may have been altered from radiolarian oozes, not from quartz sands. Shallow marine microfossils are found in some deep-sea sediments. Simple rules of thumb may not give the correct explanation of complex processes. While the natural-history approach is necessary, the method is, in some cases, insufficient, and we need to know the physics of the processes involved. What seems likely may be physically impossible. What seems improbable may be the only physically viable explanation. This is the physical-science approach to geology.

The diverse natural phenomena are infinite, and specialization of the earth sciences seems a necessity. Sedimentologists are specialists of sedimentary rocks; they are able to extract valuable data on earth history from the kaleidoscopic sedimentary record. Physical principles are, however, few in number and they are taught to geology students. Recongnizing that earth phenomena are physical processes, a person using the physical-science approach can apply the three laws of motion, three laws of thermodynamics, the principles of the conservation of energy and of matter, and perhaps a few other fundamental relations in science, to explain a diversity of geologic processes.

M. King Hubbert advocated and exemplifies the efficacy of the physicalscience approach in geology. Hubbert was a physics student in college, and may not have taken any undergraduate course in hydrology, petroleum geology, or tectonics. Yet, he made his greatest scientific contributions in these three fields of specialization: in hydrology with this theory of groundwater motion, in petroleum geology with his research on the oilproduction technique of hydraulic fracturing, and in tectonics with his analysis of the role of pore pressure in overthrust mechanics. Similarly, the concept of platetectonics, the new paradigm of geology, is innovated by students of geophysics who knew little geology and still less tectonics; they made a simple assumption that diverse earth phenomena are the physical consequences of moving rigid plates on the surface of the earth.

As I have indicated, my purpose in teaching this course is less to teach sedimentology, but rather to instill in students the skill to view earth phenomena as physical processes. In fact, my original aim was to write on the *Physical Principles of Geology*, and this was the title of the book which I was contracted in 1971 to write. After 2 decades, I see my limitations and elect only to write on those *Principles* as illustrated by sedimentary processes.

I have tried to avoid the authoritarian attitude in writing this textbook, although I realize that we cannot start out from kindergarden. Since my students are mostly freshmen or sophomores in college, I made middleschool physical sciences a prerequisite for my course, and I presumed that they have a rudimentary knowledge of differential and integral calculus. I expect the same for my readers.

Textbooks are commonly not readable. After having written three books on geology for general readers, I find no reason why should I not also use the same "trade-book" style to compose a textbook for geology students. Several of my colleagues and students have read a draft of the manuscript and they found the claim of my subtitle - '*A Readable Textbook*' - justified.

I am indebted to many persons in having produced this volume, but I could single out only a few in this short acknowledgment. M. King Hubbert was the inspiration of my teaching philosophy. Several friends, especially Max Carman, Gerald Middleton, and Harvey Blatt gave me encouragement. Dave Kersey consented to be a co-author when I first began to write, but he had to back out because of his other commitments. My former assistants, Helmut Weissert, Guy Lister, David Hollander, Ulrich Henken-Mellies and Jon Dobson, helped in many ways in my teaching of the course, and their numerous corrections of the earlier drafts helped improve the manuscript.

I owe a special thanks to Ueli Briegel who not only instructed me in the use of the word-processor, but also consented to make a laser-print copy of the manuscript for photo-offset production. His effort preserves the aesthetics of the volume and at the same time reduces the cost to potential readers. Albert Uhr and Urs Gerber prepared the illustrations. I am indebted to the many colleagues and publishers who gave me permission for reproducing modified versions of their original illustrations.

The book is dedicated to the memory of my first wife Ruth. She grew up in the land where the word *Heimweh* originated. I promised her that i would write a textbook, so that I would become known, so that I might be offered a job in Switzerland, so that she could return to her native country. It has not turned out that way. Her ashes went to Basel first, before I was called to Zurich, before I became established, before this book was written. Life is full of its little ironies, as my favorite writer Thomas Hardy would say.

Zürich, Summer 1989   Kenneth J. Hsü

Contents

1 Introduction

Why lecturing? Why this Textbook? Why Physical Principles?
Why a Readable Textbook?

I was a student once and I often thought that lectures were a waste of time. Some lectures are probably a waste of time, when students could use the same time getting more out of studying on his own. Yet, lecturing is still the *modus operandi* of instruction, from primary schools to universities. It has been time tested universally, and its value should not be underestimated. Furthermore, we teachers are hired to give lectures; we have little choice. I have, therefore, had to give much thought to the possibility of optimizing. Why would you want to attend my lectures? What can you not learn from reading a book or an article in a library?

I would like to see you here not because you have to be. As university students, you have a freedom of choice. You come, because you can learn from my lectures what you cannot in reading a book.

Students go to lectures, because textbooks are seldom readable. Few authors are used to writing in conversational style, and hardly any textbooks of science are published in that way. Richard Feynman, a noted physicist 20 years ago wrote a readable text on hydrodynamics, but he already had his Nobel Prize. Most authors are, however, required to be explicit, succinct, and, worst of all, comprehensive.

Comprehensive books serve as ready reference for teachers; they can choose what they want to learn from the encyclopedic coverage of comprehensive books. But textbooks are often nightmares for students; they do not know where they should start (and where they should end) in a big, thick textbook. Lectures have an advantage because they can never be truly comprehensive. A course has to be taught in so many hours. Constrained by time, a lecturer has to choose his materials. Attending lectures, students are told what is important and what is less important in a discipline where knowledge is unlimited. Students in our university, therefore, demand lecture notes from their instructors. These notes define a more limited area for candidates to prepare for their examinations. This book, like many other textbooks, is an outgrowth of my lecture notes. I have, however, not written this book because the lecture notes were there. I have the more ambitious aim of presenting my teaching philosophy for the consideration of other instructors. Perhaps some changes could be made in the way we teach sedimentology to our undergraduates. Perhaps some changes could be

1

made in university teaching. Finally, I hope this unusual opus, if proven successful, might start a new trend in writing textbooks.

Books are seldom written with regard to a reader's current qualifications. If he can not understand now, he can come back tomorrow, or next year. Lecturers are, however, not supposed to ignore a student's capacity to learn. If they do, they would hear complaints. After 20 years of teaching and having a reputation for being a poor lecturer, I can finally appreciate the difference between what I should teach and what students could learn. I can assure you that it is not my purpose to teach you all that I know about sedimentology. I give lectures because I hope you might learn something which you should and which you could.

You attend lectures because you do not always understand what you read. Often you may have difficulty to fathom the reasoning behind an author's statement. I remember my own frustrations as a student, especially when I had to read an article or a book involving the application of physics or mathematics. I was stopped because I encountered a technical term which I did not understand, or because I came across an elegant equation, which *is well known*, or which *can be proved*. Well, the term was probably defined in a technical dictionary which I did not have. The equation might be well known to an expert or easily proven by the knowledgable, but it was not known to me or it could not be proved by average users of a textbook.

When you face such a problem, you either waste many hours puzzling over a small point, or you give up and throw the book aside. You cannot face the author and expect an immediate answer to a question. Lecturing is different. Some professors do get away with bluffing, and claim that "it is well known", or "it can be proved". One could be stopped, though, someday by a bright young kid who calls his bluff. Any teacher with pride and sensitivity would thus not want to get into this embarassing situation. He has to be prepared to answer questions on any subject he brings up, and he has to know what he is talking about. Ultimately, no teacher can hide behind semantics in a lecture hall, and no student needs to be too shy to ask questions and to get to the bottom of things. This is the way learning should be. This is my intention in writing this book. I shall presume that you have learned middle-school mathematics and natural sciences and are taking courses at the university on those subjects; I shall start from there to discuss some physical principles of sedimentology.

Lecturing may be a monologue, but a thoughtful lecturer does try to interact with his audience. He will go back and repeat if he sees an abundance of puzzling expressions among his audience, or he will make a long story short when he sees them bored. One communicates with just so much repetition or omission as necessary. In contrast, an author pounds away on his typewriter. He tries to anticipate his readers, but he can never be sure.

We have to realize that lectures are tailor-made to students. If tailor-made clothes do not fit, the mistfit is irrevocable. Generations of my students have complained of my lectures, which have been specially designed for them. Yes, the designer has often misjudged the size of his customers, and I have had to change the content and methodology of my teaching. Authors of textbooks do not labour under such constraints.

Textbooks are not tailor-made to a particular group of readers. They have excesses and deficiencies, and I do not expect to produce a new edition to fulfil the wishes of any particular group of readers. Textbooks serve useful purposes because of that imperfection. A lecturer can use a textbook as a basis for his lectures; he can decide to subtract or add materials from what is written. A student can use this textbook as one of the books he chooses to teach himself sedimentology. He cannot expect to learn all he needs to know from one textbook. But if he could at least learn something from it, the book has justified its existence.

I have been giving lectures in sedimentology for 20 years. I have spent much time each year writing new lectures for each new class, because I usually forgot what I had taught the year before. Eventually, I realized that there is a "hardcore" which I had been teaching class after class for more than 20 years. I do not have to start from scratch again each year, if I write the basic principles down. Similarly a student can look up a subject in this book, many years after he has listened to my lecture; he does not have to learn the matter from scratch again from the incomplete lecture notes, which he may or may not have taken.

Lectures are transient yet final. Books are permanent, yet transient. These may sound like inscrutable quotes from Laotze's *Tao-te Chin*, but the fact remains that mistakes can be corrected if they are written down, and if they are read by the knowledgable. I realized that when I went over some notes taken of my lectures by students; errors which I made while writing on a blackboard were copied down and, in many cases," immortalized". There were other mistakes, they were made either because I did not master the subject matter or because they did not understand my explanation. Thus errors and mistakes made during a transient lecture are mostly irrevocable. Now that I have to write down what I said, I have an opportunity to correct error and remedy mistakes; I also can think over again some of the problems which I did not explain very well during my lectures and try to do better. If this printing or edition cannot do the job, there could be a second printing or a revised edition. What was said is said. What was written can always be changed.

A main reason which prompted me to write the book is the hope that my philosophy of teaching geology will survive my retirement 5 years from now. I often heard complaints of geology majors that they see no reason why they should take the many courses in physics, chemistry, and mathematics which are required of them. Unless they become geophysicists, they seem to need little chemistry, less physics, and almost no high mathematics at all in their professional practice. For a student who is to become an oil company employee, a course in sedimentology on facies models will do. He should be quite capable of reconstructing a depositional environment on the basis of facies analysis, and he sees little reason why he should have been subjected to examinations in Newtonian physics or Gibbsian thermodynamics.

This brings me to my comparison of sedimentology to art history. I wrote in my 1983 book for lay readers, *The Mediterranean was a Desert* :

Sedimentologists are students of sediments; they describe and analyze sediments and sedimentary rocks. They would cut a chip off a piece of carbonate rock, grind the chip into a transparent thin-slice, and examine this under a microscope. They would crush a shale, pulverize it and bombard the powder with X-rays to determine its mineral composition. They would pound on a sandstone and shake it until the sand grains become loose enough to run through a series of sieves to analyze its size and sorting. They would dissolve an evaporite (a chemically precipitated rock) and process it through a mass spectrometer to determine isotopic ratios of various chemical elements. Their purpose is to learn more about the origin of a sediment. Is it a beach deposit, a lime mud laid down on a tidal flat, or an ocean ooze?

In some instances one does not have to go through complicated procedures nor use sophisticated instruments; one can immediately tell the genesis of a rock by the way it looks. Techniques of comparative sedimentology were developed shortly after the Second World War, and the financial backing by the oil industry contributed considerably to their success. Teams were sent out to study recent sediments in various environments: river sediments on coastal plains, deltaic sediments at the mouths of major streams, marine sediments on open shelves, oceanic sediments on abyssal plains, and so on. Distinguished features were defined, then described as "sedimentary structures", and those structures serve to characterize suites of sedimentary deposits at various places. When a core of an ancient sedimentary formation is obtained from a borehole or an oil well, one can now compare its sedimentary structures with a known standard, in much the same way an art historian identified a purported Rembrandt by comparing its composition, coloring, shading, and brush strokes with known Rembrandts. Sometimes the comparison is purely

empirical. Other times there are good theoretical reasons why a sediment should look the way it does.

During a recent trip to Holland with my family, I noted that my teenage son, Peter, who is no art historian, easily spotted a Rembrandt in the Mauritshuus after half a day in the Rijkesmuseum. Little did he realize that certifying Rembrandt is one of the most difficult tasks for an art historian, because the superficial characteristics are all too easily imitated.

Certainly, a picture-book approach can be successful to a certain extent and may lead to correct conclusions, but a deeper understanding is often required to distinguish the genuine from the imitation, the truth from the falsehood.

We need physics, because geologic processes are physical processes. Superficial appearances may help separate the probable from the improbable, but geology is science and scientific truth is approached through a discrimination of the possible from the impossible. Principles of physics determine what is physically impossible. A lacustrine turbidite may look like a varve, but it is not a varve, and physical law may help us to discriminate a true varve from a false one.

A deeper reason to study physics is to learn precision in thinking. Swiss educators insist that Latin should be taught in middle school because one learns not only a language, but also the logic of a language. Similarly, my own experience has suggested to me that a professor in geophysics, David Griggs, saved me from the dangerous habit of fuzzy thinking.

In 1950, I went to Los Angeles to study rock mechanics with Griggs. On his door, was a quote from Lord Kelvin:

I often say that when you can measure what you are speaking about, and express it in numbers, you know something about it; but when you cannot express it in numbers, your knowledge is of a meagre and unsatisfactory kind; it may be the beginning of knowledge, but you have scarcely, in your thoughts, advanced to the stage of Science, whatever the matter may be.

We see descriptive terms such as steep slope, slow speed, poor sorting, etc., in sedimentology. The adjectives steep, slow, or poor are meaningless, unless we have some numerical values of the slope, of the speed, or of the sorting to compare to the numerical values of the angle of repose, of the critical Reynolds number, or of standard deviation in phi classes of well-sorted sand. To obtain numerical solutions of complex variables, one needs equations, and those equations have been derived from the consideration of equilibrium (forces, chemical potential, etc.), of conservation (momentum, energy, etc.), and of other physical principles.

INTRODUCTION

Several excellent textbooks on *Sedimentology* have appeared during the last 2 decades, including some genuine efforts to relate sedimentology to elementary physics. Yet, the usual authoritarian approach is adopted. I could, for example, recognize two categories of textbooks on the basis of their discussion of settling velocity. In most texts, old and new, the velocity was simply stated, citing Stokes' Law:

$$u = \frac{1}{18} \frac{(\rho_s - \rho_f)\, g\, D^2}{\eta} \ . \tag{1.1}$$

Two new texts, one elementary and the other advanced, went so far as to explain that the law is derived from a consideration of equilibrium of forces, and that the fluid resistance is

$$F = 6\pi \cdot \eta \cdot \mu \cdot D/2 \ . \tag{1.2}$$

Yet the reader is faced with the same frustration of having a "well-known formula" thrown at him — well known perhaps in its form but not in its content. He is told the same old story that the relation can be proved, but it is not proved. He is left with the same uncertainty as he is told that this formula is applicable if viscous resistance is dominant, but he does not know the meaning of dominant viscous resistance.

In my lectures, and now, in this book, I have attempted to go back to the basic principles of physics, to Newtonian mechanics, and to Gibbsian thermodynamics for a derivation of the quantitative relations commonly cited in sedimentology. Only a thorough understanding of their origin could explain the limitation of the validity of the many equations familiar to us in sedimentology.

This book is written as a history of inquiry, especially my history of inquiry. Natural laws are not stated as self-evident truth, but as approximations of quantitative relations, deduced from imperfect experiments. The purpose is not so much to inform, but to invite inquiries to reduce the element of falsehood of our understanding.

Supported by a grant from the Guggenheim Foundation, I spent 6 months on this project in 1972, while I was taking a sabbatical leave at the Scripps Institution of Oceanography. The sabbatical ended before my book was written. I continued the effort in 1973, 1974, and 1975. Still incomplete, I was not happy with what I had written and the manuscript was shelved for more than 12 years. I started to work on it again in the spring of 1988, when I finally seemed to find an echo from my students. With their encouragement, I decided to carry the task to its completion.

My philosophy of teaching is expressed in the final words of this introduction. I often said that one should teach science like he teaches a foreign language. Students learn a foreign language by studying its grammar and acquiring enough of a vocabulary to read, speak, and write. The grammar of the language of science is scientific logic, and the vocabulary consists of the technical expressions, which are commonly abbreviations of scientific concepts. You learn to read, speak and write science by knowing the principles and terminology; you do not need to memorize all the information contained in an encyclopedia of science.

Geologists study the history of the Earth, and sedimentary rocks are an archive of the Earth's history. The books in the archive are, for the uninitiated, books written in a foreign language. I have no intention to summarize all the books in the archive. The purpose of this opus is to record my effort in teaching my students the skill to read those foreign-language books in the archive.

The title of the book is *Physical Principles of Sedimentology*. It bears some resemblance to the title of a 1970 book by John Allen: *Physical Processes of Sedimentation*. The difference is significant. Whereas Allen attempts to use physical principles to interpret all major physical processes of sedimentation, I shall attempt to elucidate all relevant physical principles involved in interpreting sedimentology with a few processes elucidated in illustrations. The derivation of each of the principles, such as Stokes' Law, Reynolds' criterion for turbulence, Chezy's equation, Shield's diagram, Hjulstrom's curve, Bernoulli's principle, Darcy's equation, mass-action law, Gibbs criterion of chemical equilibrium, etc., will be given.

Teaching, for me, is not primarily aimed at the transmission of information. Information is transmitted only to serve the purpose of illustrating analytical methodology, logic, and principles. For such illustrations, I have relied heavily on my own first-hand experience in geologic research. In doing this, I am following in the footsteps of Konrad Krauskopf when he wrote his textbook on geochemistry. If I have given my own publications a seemingly undue emphasis, it is not motivated by vanity, but by necessity since I have to teach what I know, and I know best that on which I have done research. I would like to make it clear that this is not a book for you if you expect a comprehensive treatment. Some important sedimentological processes are not mentioned, not because they are unimportant, but because this book is not intended as a textbook of comprehensive knowledge. Important sedimentological processes which do not illustrate the physical principles discussed in this textbook are taught in a second course which used the natural-history approach to study sedimentology, for which numerous textbooks are available on the market. The emphasis of this textbook is placed on the understanding of the fundamental principles, not on the knowledge of particular processes.

For brevity, the expression *chemical* did not appear in the title of my book. Chemical processes are important in sedimentology, but they are, strictly speaking, physical processes. Whereas the latter consider mainly the mechanical energy and work involved, the former take into consideration the heat and chemical energies as well. The basic axioms, such as conservation of matter and of energy, are the same.

The subtitle of the book *A Readable Textbook of Sedimentology*, is an advertisement of my unusual approach in writing this textbook. After having written three "trade books", or books on geology for general readers, I have learned a few things about writing a readable text. One of those is that illustrations disrupt the continuity of a readable text: A picture may be better than 10 000 words, but the art of writing is to use a few well chosen words in place of a picture. I often wonder if Darwin's success with his *Origin of Species* may be traced to the fact that the book is readable because it is virtually devoid of illustrations.

A readable book avoids subheadings, but I have given running titles to facilitate quick references. The book has to be sufficiently organized so that portions of the text can stand alone and be read without a great demand on the memory capacity of the reader. A readable book can be started from any page, and be read forward or "backward", on odd occasions, or in one sitting. I have "peppered" the text with anecdotes and autobiographical vignettes, in an effort to keep up interest, especially when discourses of physics are weighing a reader down. This style of writing has been called "change of pace", as I was taught by my editors. Finally, a readable book has plenty of one-syllable words.

The author of a trade book, as my editor told me, has to make the subject matter comprehensible for the readers; he has less need to convince them. The author of a scientific article, on the other hand, needs to be convincing; comprehensibility is of secondary importance. The author of a readable textbook, however, has to be concerned with comprehensibility and credibility. Facts and figures have to be presented and some illustrations are indispensable. I have, therefore, chosen a minimum for each chapter. The fact that numerous figures have been found necessary is perhaps an admission that I have not fulfilled my promise of readability. I am, in fact, not adamant on the matter whether I should have more or less illustrations, and I would revise my opus accordingly for a second edition, if I hear very loud complaints.

Undergraduate textbooks commonly do not cite references, because all the matters discussed are supposedly common knowledge. I shall refer to the authors and shall cite their work in the section "Suggested Reading" at the end of each chapter. I shall also, in a short readable text, call attention to those studies, which greatly helped formulate my thoughts in writing this book, while giving my reasons why and to whom the further reading could be beneficial. A list of the references, arranged alphabetically, will appear at the end of the book.

Suggested Reading

For undergraduates preparing for their examinations, no further reading is necessary; they are free to agree or disagree with my teaching philosophy. For teachers of sedimentology who may wish to adopt the textbook for their instruction, I would like to suggest that they read the several articles by M. King Hubbert on geological education.

When I joined Shell Development Company as a young post-doc, Hubbert was the general consultant of the company, a wise investment by the company to promote creative and productive achievements. When I got my first teaching job at SUNY Binghamton in 1963, I went to say farewell to Hubbert and to ask his advice. He gave me two reprints, his article on *The Place of Geophysics in a Department of Geology* (AIME Tech Publ, No 945, 1938) and his *Report on the Committee on Geologic Education of the Geological Society of America* (Interim Proc Geol Soc Am, 1949). Hubbert emphasized in 1938 that the phenomena of the earth studied by geologists are also phenomena of physics. He was not as arrogant as Lord Kelvin who claimed that there was stamp-collecting and there was science, which was physics. Hubbert did point out, however, in 1949 that the natural-history approach to geology, "wherein, with but minor recourse to the relationships established in other sciences, students are trained in the syntheses that can be made from direct geological observations, .." is a necessary approach but insufficient.

Anticipating the Earth Science Revolution of the 1960s, Hubbert and his committee advocated the "physical science approach" in geology. The most effective way of geological education, in their opinion, is **"that at all instructional levels ... only those inferences be presented to students for which the essential observational data and logical steps leading to the inferences have also been presented."** Hubbert deplored the practice of **"many widely used textbooks"**, which **"have lost sight of our intellectual foundations and ... have reverted to authoritarianism"**.

In his Presidential Address of 1962 (Geol Soc Am Bull, 74: 365-378), Hubbert asked: *"Are we retrogressing in Science?"* He gave examples to illustrate that many of the propositions stated in textbooks are in fact incorrect, and that there were no valid derivations given, and propositions were "taken as true because the book said so". He further pointed out that in the whole field of science the master generalizations are few; "these include the three Newtonian Laws of Motion and the Law of Universal Gravitation, the three Laws of Thermodynamics, the two Maxwellian Laws of Electromagnetism, the Law of Conservation of Matter, and the atomic and molecular nature of chemical elements and their compounds." "These great generalizations encompass the whole domain of matter of energy — the whole domain of observable phenomena — that a modern scientist cannot afford to be ignorant of them. If he does have this type of knowledge, it is no longer necessary for him to burden his mind unduly with the infinity of details in whatever domain of phenomena he may choose to work."

Hubbert lamented the modern tendency of retrogressing in science and appealed to us that we should revert the trend to make it "mandatory for students to receive a working knowledge of all the fundamental principles of science."

I have followed Hubbert's advice during the whole of my teaching career, and this is the goal of this textbook.

2 Sorting Out and Mixing

Bulk Minerals - Heavy Minerals - Size and Sorting
Exclusion Principle

Sedimentary particles come mainly from the lithosphere, the hydrosphere, and the biosphere, and they are detrital, chemical, and biogenic in origin.

All rocks in the lithosphere could yield sedimentary detritus. Where mechanical erosion is dominant, sedimentary particles are rock fragments, also called *lithic* grains. Fine-grained rocks, such as chert and andesite, make good rock fragments. Sediments containing abundant lithic grains are commonly considered *immature*; the term immaturity refers to the fact that the detritus has not undergone extensive "wear and tear". A rock consisting of chert detritus records, however, a very different history from the rock consisting of andesite detritus. Chert-pebble conglomerates are not considered immature. In fact, the presence of chert grains to the exclusion of other more easily weathered grains testifies to intense chemical weathering in the source terrane. Only rock fragments consisting mainly of easily weathered minerals, such as feldspars, make immature sediments.

Perfectly *mature* detrital sediments are quartz sandstones, which are made up almost exclusively of quartz. Some have been derived from older quartz-rich sandstones; they are *polycyclic,* referring to the "reincarnation" of old detritus in newer sediments. Others owe their origin to the elimination of less resistant minerals in source areas, during transport, and /or after deposition.

Francis Pettijohn suggested the idea that the relative abundance of feldspars(F) and quartz(Q) is a *maturity index*. Igneous rocks have more feldspar than quartz, and their ratio F/(F+Q) is more than 50. Most sedimentary rocks, with their detritus derived from weathered terranes, have a feldspar content considerably less than 50. Feldspar-rich sandstones are immature. They are called *arkoses*, if their F/(F+Q) is greater than 20 or 25; those containing less feldspars are *feldspathic* sandstones.

Almost a century ago, an English geologist William Mackie studied the percentage of feldspar in river and beach sands. He found a downstream decrease in sands of River Findhorn (England). The percentage of feldspar is at

Delsie Bridge	42%
Logie Bridge	31%
Fornes	21%
Culbin at sea	8%

This dramatic decrease within a short distance downstream was once considered evidence of elimination of feldspar by abrasion during river transport. The deduction seemed logical and reasonable. Another explanation is that the decrease of the feldspar content has resulted from dilution by quartz-rich sands transported down the tributaries.

Several sedimentologists investigated the problem during the following decades, and the final verdict was handed down by R. Dana Russell in 1937, when he made a systematic study of the mineralogy and texture of the Mississippi River sediments. His results are illustrated in Fig. 2.1. The decrease of the feldspar content in Mississippi River sands down the more than 1700-km-long course is trivial, and this decrease can be explained by the dilution effect. Russell concluded, therefore, that the selective destruction of less resistant minerals during river transport is insignificant. Quartz, the most persistant detrital mineral, is common in mature sands not because of exceptional resistance to abrasion during transport, but because of susceptibility of other minerals to chemical weathering in source terranes.

So the pounding and grinding during river transport did little to eliminate the less resistant minerals such as feldspars. What was then the effect of wave actions on a beach. When I left the university in 1954, my first professional job was to compare the mineral composition of beach sands at various parts of the Gulf Coast, in order to evaluate the influence of abrasion during marine transport. This was my first introduction to process-oriented geology. The results of my investigations are shown in Fig. 2.2.

A first glance seems to support the postulate of feldspar destruction during longshore transport. The beach sands at the mouth of the Mississippi Delta have about the same composition as river sands of the lower Mississippi, and the feldspar content decreases westward, in the direction of transport. Such a decrease could be attributed to mechanical attrition. A close analysis revealed, however, that the decrease of the feldspar content, as expressed by the $F/(F+Q)$ value, is a manifestation of dilution. By analyzing the composition of the various size fractions of the Texas beach sands, it was found that the coarser fractions, having been derived from the Mississippi, have about the same feldspar percentage as that of the sand in the delta area, only the finer fractions, having been derived from the Tertiary quartz sandstones and carried down by small coastal streams of Louisiana and Texas, are deficient in feldspar. This precise analysis has revealed that the westward decrease of $F/(F+Q)$ value in beach sands is not caused by attrition, but is a manifestation of dilution.

The postulate that the mineral composition of sands is little changed during transport suggests that chemical weathering must have played a significant role. This idea is best illustrated by the results of studying the river and beach sands of Florida: Detritus derived from source terranes where the climate is favorable for deep chemical weathering is mature and quartz-rich. The sands of West Florida beaches, for example, are derived from the

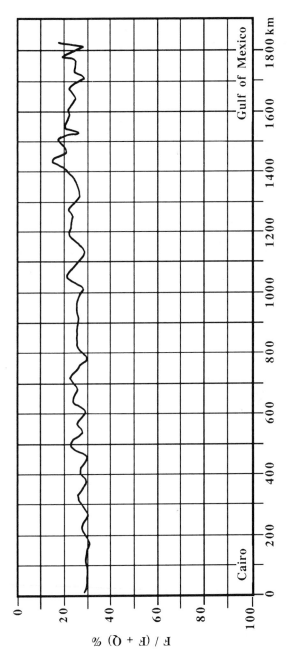

Fig. 2.1. Downstream variation of the feldspar content in the Mississippi River sands (modified after Russell 1937). The feldspar content of the river sands has changed very little, from about 30% to about 25%, in the 1700-km plus course of transport from Cairo, Illinois to the river mouth. The slight variation has been explained by the dilution effect, when the sands from the Upper Mississippi are mixed with those poor in feldspar derived from the coastal tributaries of the Mississippi. This study by R. Dana Russell falsifies the assumption that the mineral composition of a sand is modified by abrasion during river transport

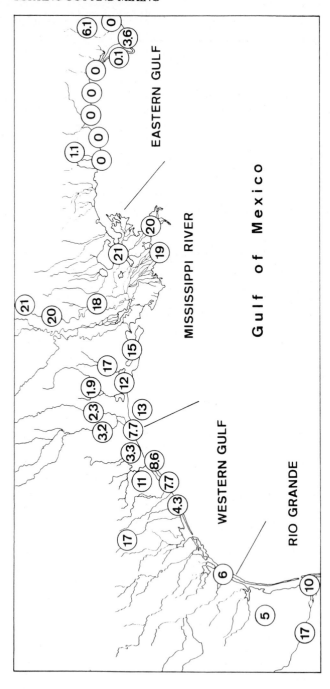

Fig. 2.2. Variation of feldspar content in the beach sands of the Gulf Coast (modified after Hsü 1964). The beach sands in the Mississippi Delta region have a feldspar content of about 20%, virtually the same as that in the river sands of the region. The feldspar content in beach sands decreases westward to a minimum of less than 5% in southwest Texas. A systematic study by Hsü in 1964 indicated, however, that the decrease is not caused by mechanical abrasion during transport, but is a manifestation of mixing of sands from different sources

weathered terranes of the southern Appalachian Mountains. The sand grains are still very angular, a testimony of their "youthful" history, having just been swept away from the red-earth hilly country of Alabama. Yet the sand consists almost exclusively of quartz, because feldspars have been weathered in the source areas and are converted into clay minerals.

We now believe that the composition of a detrital sediment is a fairly good indicator of past climate. Quartz sandstones are not uncommon in Switzerland; those of Eocene age are present in the Jura Mountains. Could we postulate that the Eocene climate was warm and humid in Switzerland so that feldspars were eliminated by chemical weathering in the source terranes? Yes, we can, and this postulate is confirmed by fossil fauna and flora evidence.

Another component of a sandstone consists in minerals denser than quartz, and they have been referred to as *heavy minerals*. Heavy is, of course, the wrong word; those minerals are not heavier, but denser than quartz or feldspar. Heavy minerals are minor constituents of sandstones, constituting commonly less than 1% of the bulk, but the presence of various species could yield significant information.

Individual heavy minerals are separated from a concentrate of sand, after the lighter particles (quartz, feldspar, etc.) have been removed by flotation on heavy liquids. H. B. Milner was one of the pioneers to use binocular microscope to identify those minerals. This kind of study was once very fashionable. My father-in-law, Hermann Eugster, remembered his post-doctoral sojourn to the Imperial College of London in the early 1920s, to study with H. B. Milner, the author of *The Principles and Practice of Correlation of Sediments by Petrographic Methods.*

Milner and his colleagues discovered a *heavy-mineral zonation* in sedimentary sequences. Pyroxenes are commonly found only in Recent sediments, amphiboles in young, near-surface sandstones, the heavy-mineral assemblage in progressively deeper sedimentary formations is characterized by successive dominance of epidote, garnet, and zircon. Eventually such zonal arrangement has been identified not only from Europe and North Africa, but also in various parts of North America (see Fig.2.3).

Eugster worked for an oil company in Colombia, and the dating and correlation of geologic formations were of prime importance to exploration. Foraminiferas and other microfossils had been found in marine sediments, but their stratigraphic significance was not well known. Much hope during the 1920s was placed on the use of heavy minerals for stratigraphical correlation. The Society of Economic Paleontologists and Mineralogists, SEPM, the earliest sedimentological society, was founded in 1930, when both paleontologists and (heavy-)mineralogists served the oil industry for the purpose of making correlations.

15

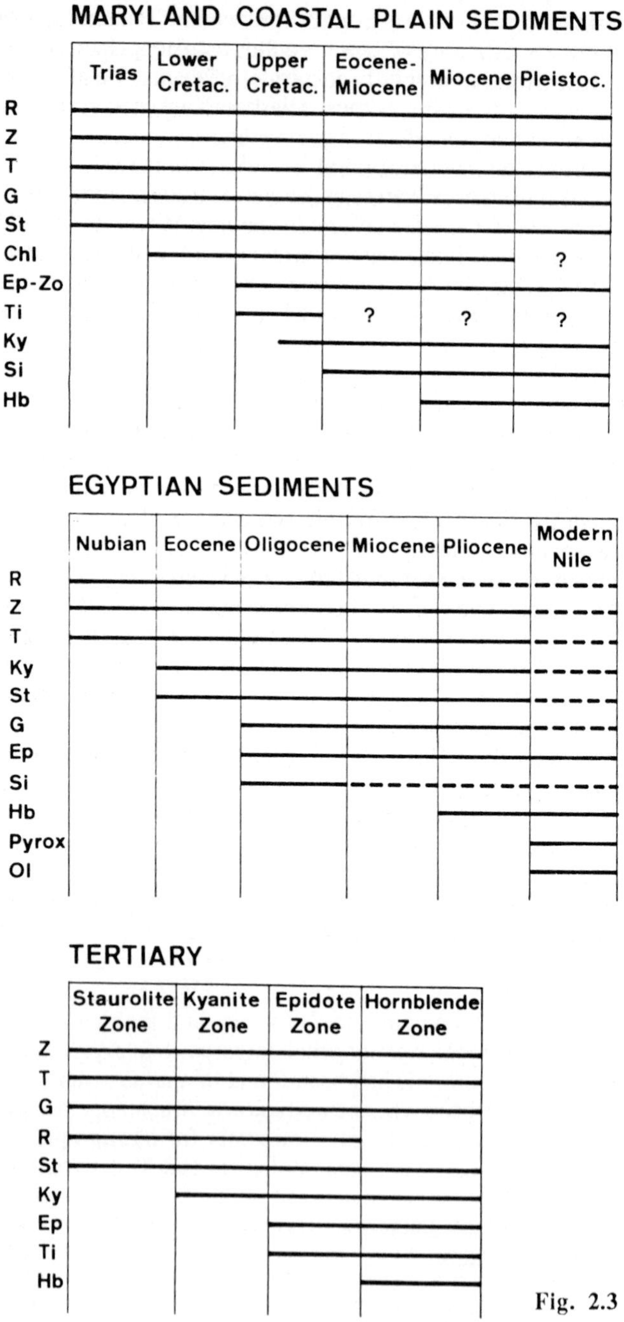

MARYLAND COASTAL PLAIN SEDIMENTS

	Trias	Lower Cretac.	Upper Cretac.	Eocene-Miocene	Miocene	Pleistoc.
R						
Z						
T						
G						
St						
Chl						?
Ep-Zo						
Ti				?	?	?
Ky						
Si						
Hb						

EGYPTIAN SEDIMENTS

	Nubian	Eocene	Oligocene	Miocene	Pliocene	Modern Nile
R						
Z						
T						
Ky						
St						
G						
Ep						
Si						
Hb						
Pyrox						
Ol						

TERTIARY

	Staurolite Zone	Kyanite Zone	Epidote Zone	Hornblende Zone
Z				
T				
G				
R				
St				
Ky				
Ep				
Ti				
Hb				

Fig. 2.3

Paul Krynine, in a series of talks given at the Annual Meeting of the Geological Society of America in 1941, was an enthusiastic proponent of the idea that the vertical zonation of heavy minerals reflects the history of erosion of source areas. The zircon suite consisting of zircon, rutile, monazite, etc., is commonly found in older sedimentary rocks. These minerals are considered polycyclic, having been derived from the sedimentary cover of the basement in the source terrane. As the cover was eroded first, these stable minerals were deposited as the heavy mineral suite (Fig. 2.4 A) of the oldest sediments in a basin. With continuing uplift, metamorphic rocks with garnet and epidote (Fig. 2.4 B) were successively eroded. The heavy mineral suite of the middle formation consists thus of a mixed assemblage. Finally the basement composed of rocks rich in amphiboles and pyroxenes is bared by erosion, so that the youngest sedimentary strata are characterized by a suite that includes all heavy minerals (Fig. 2.4 A,B,C).

The hypothesis of relating heavy mineral zonation to erosional history encouraged "economic mineralogists". At the same time "economic paleontologists" were making great strides in their correlation of microfossils. Soon friends in SEPM became enemies when the surfaces of *isochrone* (equal-age) zones determined by faunal correlation were shown by economic paleontologists to cut across those separating heavy-mineral zones. It is true that more stable minerals occur in deeper rocks and they are on the whole older, but stratigraphical correlation by mineralogical studies are not very precise. Paleontologists eventually were triumphant; many mineralogists in the petroleum industry had to abandon their heavy minerals and used the same microscopes to identify microfossils.

When I went to Houston in 1954 to work with the Shell Development Company, my boss gave me several boxes of heavy-mineral data, representing several 100 man-years of diligent work during the 1920s and 1930s. By then, the battle for stratigraphical correlation had been won by the paleontologists. But, as I found out from the files, the mineral zonation is there (Fig. 2.4), even though its stratigraphical value is questionable. This vertical zonation requires nevertheless an explanation.

Fig. 2.3. Heavy Mineral Zones (modified after Pettijohn 1957). The diagrams illustrate that rutile (**R**), zircon (**Z**), and tourmaline (**T**), the so-called stable heavy minerals, are present in all sandstones, whereas olivine (**Ol**), pyroxenes (**Pyrox**), and amphiboles (**Hb**) are present only in the youngest. Other minerals of intermediate stability include garnet (**G**), kyanite (**Ky**), sillimanite (**Si**), chlorite (**Chl**), epidote-zoisite (**Ep-Zo**), titanite (**Ti**).

Zonations, such as that in the Cenozoic formations of the Gulf Coast region, were recognized, and heavy minerals were used as "index fossils" for correlation. We now believe, however, that the zonations have resulted from dissolution of less stable heavy minerals in deeper or older rocks

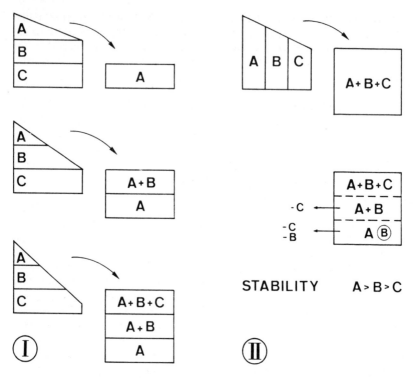

Fig. 2.4. Hypotheses of heavy-mineral zonation. Two different schools of thought have developed regarding the heavy-mineral zonation. Krynine assumed that the heavy minerals in sedimentary rocks are all stable, and the zonation is a manifestation of the changing input of source materials (*Model I*). Pettijohn postulated dissolution of less stable heavy minerals in more deeply buried rocks (*Model II*): for explanation of **A**, **B**, **C**, see text

Francis Pettijohn, among others, had noted that the heavy minerals have varied resistance to weathering or to chemical dissolution. The order of increasing resistance is pyroxenes, amphiboles, epidote, garnet, kyanite, staurolite, zircon, etc. That the zonation was caused by differential dissolution of less stable heavy minerals by groundwater was proposed.

I worked then with my late friend James Taylor to study the heavy minerals of Late Cenozoic sandstones of southern California. The bulk composition of the sandstones and the geologic history of the region are such as to indicate that the heavy mineral composition of the sands should have been the same from the top to the bottom of the sedimentary sequences. Yet, Taylor found a heavy mineral zonation, similar to that found in the Gulf Coast: the oldest suite consists of zircon and rutile, and the youngest consists of all heavy minerals. He also observed that the weight percent of the total heavy-mineral suite becomes less and less in deeper and deeper formations, and that this change can be explained if we assume the dissolution of less stable heavy minerals in deeper formations. After this exercise, I found little reason to question Pettijohn's postulate.

To draw geological conclusions on the basis of sedimentary petrology is a "natural-history approach". Syntheses are made on the basis of direct geological observations, and the recourse to the relationships established in other sciences is vague and not quantitative. We have no physical law, which states that change of mineral composition by abrasion during transport is impossible. We have made no chemical experiments to verify the dissolution of the less resistant heavy minerals. As M. King Hubbert pointed out, the natural-history approach in geology is necessary, but insufficient. The "earth-science revolution" of the 1960s came only after the "physical-science approach" had been adopted to investigate the physics and chemistry of the earth's phenomena.

Traditional approaches were conserved in central Europe, however, when I was called to ETH Zurich in 1967. Sedimentology was then taught in the form of studying heavy minerals, and only to a few students who happened to work on the Molasse formations. Thanks to a revolt by students, the course on the physical principles of sedimentology is now obligatory for our geology majors; this is the course I am now teaching. I have made, however, this lesson in sedimentary petrology my first lecture, because I do not recommend that the "natural-history approach" be abandoned. Studies of heavy minerals could yield valuable information on the provenance of detritus. For example, the discovery in Alpine Cretaceous Flysch of heavy minerals derived from ophiolites told us that the Cretaceous Tethys Ocean was bounded on the south by a coast range, where *ophiolite melanges* were exposed. We now have evidence that the ocean crust was being consumed long before the Early Tertiary continental collision which formed the Alps. A proper interpretation of sedimentary petrology has thus made an important contribution to our understanding of the geologic history of Switzerland.

We have so far been discussing separation by chemical processes; such sorting leaves an imprint on the composition of a detrital sediment. Sorting by mechanical processes determines its texture.

The grain-size distribution of a detrital sediment is commonly determined by sieve analysis. A series of sieves is used, each successive one has an opening half as large. The weight of the sedimentary particles left in each sieve constitutes the *sieve fractions* or the weight fraction of a grain-size class. The median diameter of a sediment is the average diameter, which is larger than 50% of the grains and smaller than the other 50%. The sorting of a sediment can be expressed by noting the distribution of components in various sieve fractions. If all of the detrital particles of a sediment are caught in the same sieve fraction, the sediment is well sorted. If its detrital grains are found in several sieve fractions, the sediment is not so very well, or poorly, sorted.

Standard sieves are made such that each successive sieve opening is only half as large. Sieve fractions are thus expressed as powers of 1/2. Mathematically minded sedimentologists invented phi units, and ϕ is a dimensionless number defined by

$$\phi = -\log\left(\frac{1}{2^n}\right) \quad , \tag{2.1}$$

where $(1/2^n)$ is the median diameter (in mm) of a sand, with $n = 0, 1, 2, \ldots$. A sediment with a median diameter of 1 mm (n=0) has a phi value of 0. One with a median diameter of 0.125 or $1/2^3$ mm (n=3) has a phi value of 3. A clay with a median diameter of 0.002 or $1/2^9$ mm (n=9) has a phi value of 9.

Opinions have varied on the definition of sand, silt, and clay; European and American usages are different (Table 2.1). The North American practice advocated by Udden has been adopted by many, if not most, of the sedimentologists.

Unsorted sands have nearly equal weight percentages in numerous sieve fractions. Mechanical sorting results from the fact that a river, or a marine current, has a range of speeds which could erode or transport detrital particles of certain size. Well-sorted sands have thus sand grains which almost all belong to the same phi. Such good sorting is evidence that the velocity of the current passing through the site of deposition varied within a narrow range. Coarser sediments had all been left behind before the current reached the site of deposition; finer particles remained in suspension and were carried away. Only sediments of a given size fraction were too coarse to be transported farther. In Chapter 4 this matter of sorting during sediment transport will be discussed.

Sediments of different origins can also be separated from one another because their paths of sediment transport are different. Coastal marine current, also known as longshore current, tends to flow in a certain direction so that the detritus transported by this current system would be excluded at sites in the opposite direction. On the Gulf Coast, for example, the beach sands east of the Mississippi River delta contain little feldspar (Fig. 2.2), because the detritus from the river have all been carried westward by longshore currents. The quartz sands on Florida beaches have been derived from the weathered terranes of the southeastern United States.

Table 2.1. Classifications of detrital sediments (modified after Pettijohn 1957). Detrital sediments can be classified on the basis of grain size into clay, silt, sand, gravel, etc. It is a sad commentary on the triviality of scientific endeavor when scientists found little better to do than to argue if a clay-sized particle is less than 0.001, 0.002, or 1/256 mm.

Despite the difference in opinions, the one central feature common to all the classifications is the fact that size classes are geometrical progressions. Such schemes have been proposed because of their practicality. I have analyzed the grain-size distribution of unsorted sediments, and found that the weight percentages of all sieve fractions, with the size of sieve openings in geometrical progression, are the same for a perfectly random grain-size distribution. Unsorted sediments, in the new mathematic language of Mandelbrot (1982), have a fractal distribution of grain size

	HOPKINS 1889		ATTERBERG 1903		UDDEN 1914		WENTWORTH 1922		CAILLEUX 1929	US BUREAU OF SOILS	
1000			Block		Boulders		Boulders				512
						vc			Blocs		256
						c		Cobbles			128
100			Sten		Boulders	f			Galets		64
						vf					32
						vc					16
10			Grus		Gravel	c		Pebbles	Graviers		8
						f					4
						vf		Granules		Gravel	2
1.0	Gravel		Sand		Sand	vc	Sand	vc			1
	Sand	c				c		c	Sables	c	1/2
						f		m		m	1/4
		m	Mo	Fimma		vf		f		f	1/8
0.1		f			Silt	vc		vf		vf	1/16
		c		Mjäla		c				Silt	1/32
0.01	Silt	m	Lättler	Vesa		f		Silt	Poussières et Boues		1/64
				Mjuna		vf					1/128
		f			Clay	vc				Clay	1/256
0.001			Ler			c		Clay			1/512
	Clay					f					1/1024
						vf					1/2048

21

Exclusion could be a consequence of geographic isolation. The sands on Bahama Islands, for example, have neither feldspar nor quartz; the Florida Strait has served as a barrier to "filter out" all terrigenous detritus.

Where the *siliciclastics,* i.e. silica and silicates, are "filtered" out, the streams draining a carbonate terrane could carry down limestone and dolomite debris to marine realms to form carbonate sediments, such as lime sands or *calcarenites* .

More commonly, however, calcium carbonate sediments are derived from skeletal debris of organisms. The Recent carbonate sediments on the shallow marine Great Bahama Bank consist mainly of fossil skeletons or debris. Oozes on the bottom of open oceans are also mainly biogenic. They are derived from skeletons of nannoplankton, foraminferas, radiolarias, and diatoms; the detritus from land have been deposited nearer to shore.

Sedimentary particles can originate from chemical precipitation, and they form chemical sediments if terrigenous detritus are excluded. Chalk sedimentation in Lake Zurich is a good example. The suspended particles from the Linth, its main tributary, have been filtered out before the river water leaves the Lake of Walenstadt, while the debris from smaller tributaries are deposited behind flood-control constructions. As a consequence, a chalk, not a marl or clay, is now deposited in Lake Zurich.

Carbonate muds in marine environments were once thought to be chemical precipitates. Work by Robert Ginsburg and others during the 1950s have shown that they are in fact mostly biochemical precipitates by green algae.

Chemical precipitates from lakes and oceans include not only carbonates, but also sulphate and chloride *evaporites* (anhydrite, gypsum, halite, etc.) as well as phosphates and other important economic minerals. Whereas the genesis of siliciclastics involves mainly physical processes, the origin of carbonates, evaporites and other monomineralic deposits are problems of biomass production and chemical reactions, which will be discussed later.

Sorting by chemical processes leaves an imprint on the mineralogical composition of a sediment, sorting by mechanical processes determines grain size. The mechanical sorting is done by physical processes, governed by physical laws. I shall start my discussion of those principles in the next chapter with Newton's First Law of Motion.

Suggested Reading

The natural-history approach in geology is inductive, and induction means "the bringing forth of facts to prove something". Philosophers of science, in their jargon, tell us that you cannot "prove", you can only falsify. In other words, no accumulation of facts suffice to prove the truth, but a single fact can prove an assumption wrong. Therefore, I have referred to two studies (Mississippi River and Gulf Coast), which falsified the assumption that the composition of a sediment is changed by abrasion during transport, and I have not tried to prove, by citing many case histories, that mineralogy indicates provenance.

Many excellent studies on sedimentary petrology have been published, and they give us an appreciation of the complexity of interpreting sediment composition and texture. Undergraduate students could begin by reading Francis Pettijohn's *Sedimentary rocks* (Harper, New York, 1957), especially chapters 2, 3, 11, and 12. Advanced students writing a term paper, or doing a dissertation on physical sedimentology may wish to dig deeper into the classic papers by William Mackie on *The sands and sandstones of eastern Moray* (Trans Edinburgh Geol Soc, 7:148-172, 1896), by R. Dana Russell on the *Mineral composition of Mississippi River sands* (Bull Geol Soc Am, 48:1306-1348, 1937), and other papers cited by Pettijohn in his book.

The major conclusions on mineralogy, as summarized in this chapter, came as a revelation to me, when I had to synthesize my data for an article on *Texture and mineralogy of the recent sands of the Gulf Coast* (J Sediment Petrol 30:380-403, 1980). I owe much to Robert Ginsburg, however, for having taught me the fundamentals of carbonate sedimentation. Much of his wisdom can be found in his article on *Environmental relationships of grain size and constituent particles in some South Florida carbonate sediments* (Bull Am Assoc Petrol Geol 40:2384-2427, 1956).

Francis Pettijohn has personified the natural-science approach to sedimentology. For those endeavoring to obtain a deep appreciation of what we have learned in sedimentology with this approach over the past 70 years, they should secure a copy of the Pettijohn Festschrift *Evolving concepts of sedimentology*, edited by Robert Ginsburg (Johns Hopkins Univ Stud Geol No 21, 1973).

3 Grains Settle

Stokes' Law - Why Derivation- Dimensional Analysis
Fluid Resistance - Reynolds Number- Resistance Coefficient
Lacustrine Varves

Our students who had to take several courses in sedimentology during their four and half years of study, were required to take one examination on this subject, usually a 10-minute oral by an examinator, who was usually the chairman of the department. Our university had retained the traditional educational system of Europe, derived from a time when professors knew his few students and examinations were a formality. Occasionally, when the ordinary examinator was on leave, I was privileged to give the examination. The students would come to me and ask how they should prepare for their sedimentology test.

What can a person ask in 10 minutes? I decided that I should elect one of two themes: Stokes' law or the origin of dolomite.

One could read in any textbook on sedimentology that Stokes' law describes the settling velocity of a particle, u :

$$u = \frac{1}{18} \cdot \frac{\rho_s - \rho_f}{\eta} \cdot g \cdot D^2 \quad , \tag{3.1}$$

where ρ_s is the density of the settling particle

ρ_f is the density of a fluid medium, through which the particle settles,

η is the viscosity of the fluid medium,

g is gravitational acceleration,

D is the diameter of the settling particle.

This law is used to deduce the motion of a particle settling through a standing body of water.

My students would memorize the formula, and, armed with a calculator, they were ready to tell me the settling velocity of any sedimentary particle. They did not need their calculator, because I was not testing their computing skill under stress; I wanted, instead, to find out if they knew the basic principles of physics as applied to sedimentology.

"What is the relation of the settling velocity to grain size?" I asked.

"It varies with the square of the grain diameter".

"Is that not in conflict with the physics we learned in high school? We learned two things from Gallileo: Falling objects have no constant velocity, they accelerate. Also, the acceleration is independent of mass so that the heavier and lighter objects reach the ground simultaneously."

"The physics of grain settling is not the same as the physics of falling bodies demonstrated by Gallileo. He threw down two lead weights, and the air resistance to the falling weights is vanishingly small. The weights accelerate according to Newton's Second Law, velocity is a function of acceleration, not of mass. But the fluid resistance is not negligible when a sediment particle settles through a standing body of water."

So far so good. Then came the key question:

"Which force is greater when a grain is sinking down at its settling velocity, the weight or the fluid resistance?"

"The weight."

"Why?"

"Because the particle is sinking."

This was the type of intuitional response from my students. Philosophers before Gallileo's time thought that way too. But we now have learned the Newtonian mechanics.

Newton's First Law constitutes a first lession in middle-school physics. We learn to memorize those words:

"Every body persists in its state of rest or of uniform motion in a straight line unless it is compelled by forces impressed on it."

Stokes' Law is an expression of this First Law. When a particle sinks at a constant velocity, the two forces, gravity F_g and fluid resistance F_r, are equal and opposite. My student failed his test!

A few years later, I had to give an examination again. The student was again told that I would be satisfied if he knew Stokes' Law. He came prepared and did fine until he invoked Newton's First Law.

"Where did the resistance come from?"

"The resistance comes from the pressure on the sphere as it sinks."

"What is pressure?"

I anticipated a simple definition that pressure (P) is force (F) per unit area (A). The student, however, blurted out an irrelevant answer:

"Pressure is a function of temperature."

"Is this the definition? What is the temperature dependency of pressure?"

He did not answer my question. He drew a pressure-temperature phase diagram of a solid-solid reaction which he learned from metamorphic petrology. The sketch was supposed to support his contention that pressure is a function of temperature.

"But, I am asking you the definition of pressure, the physical entity, not its variability. What is the relation of pressure to force?"

Instead of receiving a simple answer, I was given a lecture.

"Professor Hsü, this is an examination on sedimentology, not on physics. I passed my physics examination when I graduated from middle school, and again when I took college physics. I do not have to be examined in physics again. You told us to learn Stokes' Law, and I know that by heart. For a sedimentology problem, we only need to know how long it takes for a grain of certain size to settle. It is unfair that you ask me the definition of pressure."

So he could have his way, I asked him to calculate the settling velocity of a clay particle of 0.001 mm in size, giving the viscosity of water as 1 centipoise (10^{-2} g cm^{-1}s^{-1}). With his HP calculator, he grinded out an answer right away:

"About 10^{-4} cm/s or about 30 m per year."

"Good. What is the settling velocity of a 0.002 mm grain?"

"4×10^{-4} cm/s or 120 m per year, just fast enough to settle down to the bottom of Lake Zurich in one year."

"Very good. What is the settling velocity of a coarse sand grain 1 mm in size?"

The answer from the HP calculator was 100 m/s. The student looked up with a puzzled expression.

"What is the settling velocity of a coarse gravel 10 cm in size?"

Now his HP told him that it should be 10 km/s, faster·than a canon shot.

The student did not fail his examination, because he had enough presence of mind to tell me that Stokes' Law is not applicable to the calculation of the settling velocity of sand and gravel. He was persuaded, at the same time, that it is fool-hardy to take any formula from a textbook and make calculations to obtain numerical answers. Before he can use Stokes' law, he has to understand the physics of the phenomenon. That means to dig into the origin and find out the derivation of the equation. Slowly he and other students began to appreciate the need to know the definition of pressure and of other entities in physics.

Where did the Stokes Law come from?

We are told, in one textbook, that **"a fundamental equation of fluid dynamics, derived from a simplification of the Navier-Stokes equations of motion, states that there must exist in fluids a balance between the local viscous and pressure forces."** Stokes' law is a solution of **"this equation of creeping motion for low Reynolds number flow past a sphere"**. No beginning student in sedimentology can be expected to understand this "foreign language". What is a viscous force ? What is a pressure force? Why should they be balanced? What are Navier-Stokes equations of motion? What are equations of motion? What is a creeping motion? What is Reynolds number? How was the solution of the equations of motion achieved? We cannot understand the two sentences in that introductory textbook if we do not understand all those strange expressions.

Another textbook used a simpler approach, we are told that the Stokes' law states the balance of moving and resisting forces. The moving force for a sinking sphere is:

$$\frac{4}{3} \pi (\rho_s - \rho_f) g \left(\frac{D}{2}\right)^3 .$$

(3.2)

The resisting force is

$$F_r = 3 \pi \eta u_s D .$$

(3.3)

I can see that the moving force is the effective weight of a sinking sphere with density ρ and diameter D (Fig. 3.1), but where did the author get the expression for the resisting force?

I consulted a third textbook on physics of sediment movement and was told that Eq. (3.3), not Eq. (3.1), is Stokes' law. The authors referred to the same Navier-Stokes equations of motion, and I learned that the **"difficult partial differential equations cannot be solved, except...."** Stokes, by making some assumptions and neglecting some terms, did achieve the impossible; he solved the equations by **"integrating viscous and pressure forces over the entire surface of a sphere"** and arrived at Eq. (3.3).

That did not help me. I took out a textbook on hydrodynamics and found the citation of Stokes' paper, and went out to the library to check out that volume of the proceedings of the Royal Society. I even struggled through the classic paper, and tried to present the derivation by Stokes to my students in sedimentology.

That was the time when I had just began teaching, and the experience was a catastrophe. I could not get my explanation across without giving a lesson on partial differential equations, and I was not a particularly good mathematics professor. My students were totally confused. I myself did not quite understand the physics of the problem, even though I could follow Stokes' mathematic solution step by step.

After 25 years of teaching experience, I began to appreciate the difference between scientific research, engineering practice, and teaching. In the search for truth, precision is required, no matter how difficult and time-consuming it is to achieve the final result. In engineering practice, empirical relations are handy references to get quick results, which, in most cases, satisfy practical needs. Teaching is hardest, and I have been searching for a synthesis between the rigorousness of science, as represented by high mathematics, and the easy comprehensibility of engineering practice. In more recent years, I stopped trying to teach my students Navier-Stokes equations. In hydraulic engineering, Stokes' law is presented as an

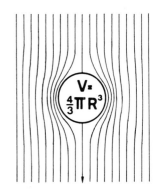

Fig. 3.1. Equilibrium of forces of a sinking sphere. A sphere sinks in a fluid under its own effective weight, which is its effective density times its volume times the gravitational acceleration. Stokes' law is derived from the consideration that the weight is equal and opposite to fluid resistance which is related to the settling velocity. What is this fluid resistance? The search for a formulation of the fluid resistance is the main emphasis in this chapter

experimental relation, an empirical law, valid for a limited range of conditions. I have, therefore, adopted this empirical approach in teaching the physical principles of sedimenology to my geology students. This is also the approach for writing this book.

Hydrodynamics deals with the mechanics of fluid movement. The motions of single, rigid bodies can be evaluated by simple exact relations such as the three Newtonian Laws. A fluid has countless molecules, and these move in various directions at different speeds, while encountering varying resistance. Consideration of physics can work out the functions of the various variables, but rarely the exact numerical relations. Numbers in engineering formulae are parameters of the statistically average behavior of a mass. Exact numbers are, therefore, usually not derived from theory, but obtained by experimentation and *dimensional analysis*.

What is dimensional analysis?

Physics relates the numerous variables in natural phenomena. The common variables in mechanics are geometry, mass, time, velocity, acceleration, force, energy or work, power, stress and pressure, viscosity, etc. Each variable has a dimension or dimensions. The three fundamental variables are length L, mass M, and time t. All other variables in mechanics should be reducible in terms of L, M, t. We have, for example:

area	A :	L^2
volume	V :	L^3
velocity	u :	$L\,t^{-1}$
acceleration	a:	$L\,t^{-2}$
force	F :	$M\,L\,t^{-2}$
energy	E :	$M\,L^2\,t^{-2}$
power	P :	$M\,L^2\,t^{-3}$
pressure, stress	p,σ:	$M\,L^{-1}\,t^{-2}$
viscosity	η:	$M\,L^{-1}\,t^{-1}$.

These are some of the physical variables which we shall be dealing with in this course.

Dimensional analysis is a common methodology in hydraulic engineering. Its fundamental principle is that both sides of a mathematical equation must, by definition, have the same physical dimensions. Take, for example, the definition of velocity

$$u = s\,/t\;.$$

The dimension on both sides are

$$L\,t^{-1} = L\,t^{-1}\;.$$

We can express the relation between velocity, time, and distance by the expression

$$\frac{u\,t}{s} = 1\;\;.$$

Formulated in this way, we can say that the dynamics of rigid-body motion, as expressed by the ratio u·t/s, can be represented by a dimensionless number, and this number is unity.

Or, if we want to relate travel distance to acceleration and time, we could state

$$\frac{s}{a\,t^2} = \frac{1}{2} = \text{dimensionless number}$$

because

$$\frac{L}{L\,t^{-2}\,t^{+2}} = \text{dimensionless.}$$

This dimensionless number is, however, not unity, but 1/2 for a moving object starting from a stationary position and moving at constant acceleration. We can also see things from the angle of dimensional analysis, and say that the variable acceleration is related to the variables distance and time by a dimensionless number.

Is it possible to relate force to some other variables by a dimensionless number? We can start by considering the shearing force required to induce the viscous motion of a liquid. In this case, the force per unit area or the shearing stress (τ) is related to velocity by the definition of viscosity (Fig. 3.2):

$$\eta = \frac{\tau}{\left(\dfrac{du}{dy}\right)} \cdot \qquad (3.4)$$

where du/dy is the velocity gradient. By rearranging the terms, we have

$$\tau = \eta \left(\frac{du}{dy}\right) . \qquad (3.5)$$

Viscosity is not a dimensionless number. As indicated by Eq. (3.4), viscosity is a stress/velocity gradient, and its dimension is:

$$\frac{M}{L\,t} = \frac{\dfrac{M\,L\,t^{-2}}{L^2}}{L\,t^{-1}\,L^{-1}} \cdot$$

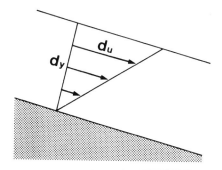

Fig. 3.2. Viscosity and velocity gradient. The velocity gradient of a viscous flow, **du/dy**, is directly proportional to shearing stress, and inversely proportional to viscosity

Substitute force in place of stress times area in Eq. (3.5), and rearrange the terms, we then have

Force = dimensionless number · viscosity · velocity · linear dimension

$$F = Z \, \eta \, u \, D \, . \tag{3.6}$$

In this and other equations, Z is a symbol for a dimensionless number; its value can be different in different equations.

This force is related to fluid viscosity, it is, therefore, called *viscous force*. Note that the viscous force is a linear function of velocity.

Is it possible to relate force to some factors other than viscosity by a dimensionless number? Yes, we can start by considering the force which imparts kinetic energy to a system. The kinetic energy is defined by

$$\mathbf{K.E.} = 1/2 \, M \, u^2 \, .$$

For a fluid of density ρ and volume V, the kinetic energy is

$$\mathbf{K.E.} = 1/2 \, \rho V \, u^2 \, .$$

The kinetic energy can also be expressed as force · distance, or stress · area · distance, or

$$\mathbf{K.E.} = F \, s = \tau \, A \, s \, .$$

A comparison of the definitions of the kinetic energy stated above indicates that the product

density · velocity square · length square

has the same dimension as the force, or the ratio

$$\frac{\rho \, u^2 \, A}{F}$$

is dimensionless. In other words, we could relate F and $\rho u^2 A$ by a dimensionless number Z, or

$$F_i = Z \frac{\rho \, u^2}{2} \, A \, . \tag{3.7}$$

This is an expression of fluid force which we see again and again in hydraulic engineering textbooks. It is called inertial force F_i, because the force overcomes the inertia of a system and imparts kinetic energy to it. What is inertia then?

We all know the inertia of daily life. To begin any project, we have to overcome our inertia. and this effort can be considered as an inertial force. A fluid at rest also has an inertia to resist movement, and inertial fluid force is that which overcomes the fluid inertia. This force, as shown by Eq. (3.7), is proportional to the square of the velocity.

Now we are ready to consider the resistance to the movement of a sinking sphere: It is a force, but is it an inertial force or a viscous force? It could be one or the other, or it could be a combination of the two and expressed by

$$F_r = f(u) + f(u^2) . \tag{3.8}$$

This equation was, in fact, formulated by Coulomb in the late 18th century to resolve the controversy whether fluid resistance is proportional to velocity or to velocity square; his Solomonic judgment was that the resistance is a function of both.

For the moment, we may circumvent the question, because a force is a force and the magnitude of a resisting force, be it viscous or inertial, can be expressed in terms of one or the other. We can, for example, express the magnitude of resisting force in terms of inertial force only, or

$$F_r = Z f(u^2) . \tag{3.9}$$

Perhaps I can emphasize this point by saying that money is money and all your assets can be expressed in terms of US dollars, even if you may own properties in foreign countries.

Let us imagine an interview between my brother-in-law Shen, who works for the Collector of the Internal Revenue Service, and a taxpayer:

"I come to ask why the rate of my property tax is 1.31 pro mil," the visitor complained, "but my neighbour, who also has assets in a foreign country enjoys a lesser rate of 1.27 pro mil".

"You have assets in Switzerland, don't you?" Shen asked. "What is their total value?"

"It is very complicated. I have cash savings, stocks and bonds, and real-estate properties. I have reported all that in my income-tax return."

"Yes, you did. Our computer figured it out, too: The value of your foreign assets amounts to 31% of that of your assets in the United States."

"That is about right"

"We charge you a tax of one pro mil for your American assets and one pro mil for your foreign assets which are 0.31 of your American assets. Therefore, your property tax rate is 1.31 pro mil of the value of your American assets. Your neighbour pays 1.27 pro mil for his property tax, because his foreign assets are only 27% of his American assets".

In other words, we can figure out a persons property tax by using the formula

Tax = dimensionless number · values in dollars of US assets.

Similarly, we can write

Fluid resistance $= Z ·$ inertial force

$$F_r = Z \frac{\rho\, u^2}{2} A .$$

This resistance is not necessarily an inertial force, but its magnitude can be expressed in such terms. We can substitute Z by another dimensionless number defined by the relation

$$ZA = C_f D^2 ,$$

where C_f is a dimensionless number called the *resistance coefficient*. We have thus:

$$F_r = C_f \frac{\rho\, u^2}{2} D^2 ,$$ (3.10)

where D has the dimension of length and is, in the case of a sinking sphere, its diameter.

Equating this resisting force to the gravity of the sinking sphere, or Eq. (3.2),

$$\frac{4}{3} \pi\ (\rho_s - \rho_f)\ g \left(\frac{D}{2}\right)^3 = C_f\ \rho_f \frac{u^2}{2}\ D^2 ,$$

rearranging the terms, we have

$$u^2 = \frac{\pi}{3C_f}\ \frac{(\rho_s - \rho_f)}{\rho_f}\ g·D .$$ (3.11)

Comparing our results to Stokes' law, we found that neither the resisting force nor the sinking velocity is represented by the same expression: Eq. (3.3) states that the fluid resistance is linearly proportional to the velocity, but Eq. (3.10) states that the force is proportional to the square of the velocity. The two formulae, Eqs. (3.1) and (3.11), for the settling velocity are correspondingly different. Furthermore, the discrepancy seems serious, because a linear phenomenon is fundamentally different from a non-linear phenomenon,

As we have mentioned, all forces have the same dimension, regardless of their nature. Whether the force is to be formulated by one or another expression, it should have the same dimensions. Dividing Eq.(3.6) by Eq.(3.10), and noting that the ratio of forces is a dimensionless number, we have:

$$C_f = Z \frac{\eta}{\rho u D} \ . \tag{3.12}$$

Since C_f and Z are both dimensionless numbers, the ratio $\eta/\rho u D$, must also be dimensionless because both sides of the equation are to have the same dimension. We can check that: you may find that this ratio on the right side of the equation, with dimensions $M L^{-1} t^{-1} / (M L^{-3})(L t^{-1})(L)$, is in fact a dimensionless number. The reverse of this ratio is, of course, also a dimensionless number, which is now called *Reynolds number* :

$$\mathbf{Re} = \frac{\rho u D}{\eta} \ . \tag{3.13}$$

A comparison of Eq. (3.12) and (3.13) indicates that the resistant coefficient of fluid flow and the Reynolds number are related, or

$$C_f = f(\mathbf{Re}) \ . \tag{3.14}$$

What is Reynolds number? Does it have a physical meaning?

Reynolds number is named after a 19th century English physicist Osborne Reynolds. The number can be considered a measure of a certain quality of fluid flow.

Every school child is familiar with numerical evaluation; he gets a grade for each course he takes. The grades may be A,B,C,D,E,F as in US institutions, or 1, 2, 3, 4, 5, 6, as in many European schools. These numbers do not have a unit; they are dimensionless. Four does not mean 4 meters, or 4 kilograms, or 4 hours. Four is simply 4, a dimensionless number, denoting the passing performance of a student in Swiss schools.

Reynolds number can be considered as a numerical evaluation of the behavior or the quality.

What quality?

Recalling that we have obtained the dimensionless ratio $\rho u D/\eta$ through a division of Eq. (3.10) by Eq. (3.6), Eq. (3.10) is an expression of inertial force and Eq. (3.6) an expression of viscous force. We have, therefore,

$$\mathbf{Re} = \text{dimensionless number} \frac{\text{inertial force}}{\text{viscous force}} \quad .$$

In other words, the Reynolds number is a numerical expression of the relative importance of the inertial and viscous forces. The numerical value of the Reynolds number is not the numerical ratio of the two forces, but a measure of the relative importance of the two.

Perhaps I can express this by using our example of grades in school again. If I use grading to judge the relative productivity of a student, this number is an expression of the ratio of his performance related to his I.Q. Receiving a grade 5 does not mean that a student's performance is 5 times his I.Q., but means that an intellegent student, with little effort, has given a good performance, or that a mediocre student, with much effort, has done better than average. Similarly, the numerical value of Reynolds number, which commonly varies up to 10^9 is a relative measure. Flows with a Reynolds number smaller than 2000, such as when you pour olive oil out of a small bottle, are sluggish and dominated by viscous forces. Flows with a larger Reynolds number, such as the flow of air past an airplane, is vigorous. Reynolds number, like the grade which you received for your chemistry examination, is a numerical measure of one aspect of the performance of fluid flow. As we shall discuss later, there are other dimensionless numbers to measure other aspects of fluid flow.

Why is the Reynolds number called by such a name? Osborne Reynolds was troubled by the two different laws of resistance to fluid flows,

$$F = f(u)$$

or

$$F = f(u^2).$$

He noted also that

"**the internal motion of water assumes one or other two broadly distinguishable forms: either the elements of fluid follow one another in lines of motion which lead in the most direct manner to their destination, or they eddy about in sinuous paths the most indirect possible.**"

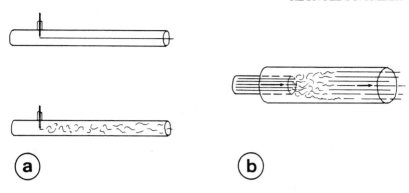

Fig. 3.3. Reynolds criterion of turbulence. **a** Reynolds visualization experiments with injected dye and glass tubes established that a laminar flow may become turbulent if the flow velocity is increased. **b** Reynolds experiments also established that a laminar flow can also become turbulent with a sudden increase of pipe diameter

We now call the former mode of motion *laminar flow*, the latter *turbulent flow* (Fig. 3.3). Intuitively, we all know turbulence is related to the velocity of the flow. In Reynolds' experiments, however, he could not pick out a particular velocity to define the transition from laminar to turbulent motion; other factors are also important. He found experimentally that the birth and development of eddies are related to (1) the velocity, (2) the diameter of pipe, and (3) the *kinematic viscosity* (ν) of the fluid, which is defined by the relation:

$$\nu = \frac{\eta}{\rho} \ .$$

The fluid flow tends to become turbulent with higher velocity, larger pipe diameter, and smaller kinematic viscosity, or:

$$\text{Potential of being turbulent} = \text{velocity} \ \frac{\text{pipe diameter}}{\text{kinematic viscosity}} \ .$$

This potential of being turbulent is dimensionless as shown by the relation:

$$\frac{(L\,t^{-1})\,L}{(M\,L^{-1}\,t^{-1})\,/\,(M\,L^{-3})} = \text{dimensionless} \ .$$

This dimensionless potential, which can also be expressed by $\rho u\, D/\eta$, is now called the Reynolds number. In other words, the potential of a fluid to become turbulent depends upon the relative importance of the inertial and viscous forces; fluid flows become turbulent when inertial forces are dominant.

The Reynolds number of fluid flow is calculated by experimentally measuring the various variables η, ρ, u, D of a fluid flow. Fluid flows with Reynolds numbers smaller than 2000 have laminar motion and the viscous force is dominant; fluid resistance is a linear function of velocity. Fluid flows with a larger Reynolds number have turbulent motion, and the inertial force is dominant; fluid resistance is a function of the velocity square. Reynolds' experiments have indicated that not only the magnitude of the fluid resistance, but also the mode of flow motion is related in some way to the Reynolds number.

Equation (3.14) states that the value of the resistance coefficient is related to the Reynolds number of fluid flow. What is this relation? How can we elucidate this relation?

The Reynolds number is calculated on the basis of measurements. The value of C_f, however, can not be directly measured, it has to be determined by experimentation.

Equation (3.11) expresses the settling velocity of a grain in terms of C_f, the grain and fluid densities, and the grain diameter. Since ρ_s, ρ_f, g and D are known quantities and the settling velocity can be determined by experimental observation, the values of C_f for sinking objects of various sizes, shape, and density, in fluids of various density, can be calculated on the basis of experimental results. These experimental values can be compared to the calculated values of the Reynolds number of fluid motions induced by sinking objects; for each experiment, the values of **Re** and C_f are determined.

Figure 3.4 illustrates graphically Eq.(3.14) relating C_f and **Re** on the basis of the experimental results. The ordinate gives the values of C_f and the abscissa those of the Reynolds number. The shape of the curve indicates:

(1) For flows with smaller **Re** values, the value of C_f is inversely proportional to **Re**, by the function

$$C_f = 24\,/\,\mathbf{Re} = 24\,\frac{\eta}{\rho D u}$$

$$(3.15)$$

and

(2) The value of C_f is about the same for flows with a large or very large Reynolds number.

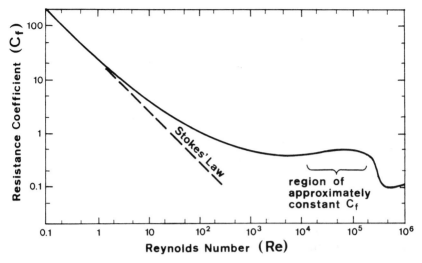

Fig. 3.4. Relation between resistant coefficient and Reynolds number (modified after Prandtl at al. 1969). The values of the resistant coefficient C_f, calculated on the basis of experimental data, are found to vary linearly with the **Re** for flows with low values of Reynolds number. This relation is that predicted by Stokes on the basis of his theoretical estimate of viscous resistance. The linear relation no longer holds true for turbulent flows with large values of Reynolds numbers

Over the years in my lectures, I had difficulty in explaining the meaning of these relations. We all think of the resistance coefficient C_f, relating resistance to velocity, like a coefficient of friction relating friction to normal force, the resistance is larger if the coefficient is larger. Yet if the value of C_f is inversely proportional to velocity, it would seem that the fluid resistance was smaller for greater velocity. In fact, of course, the value of C_f is not the most important factor in determining the total resistance, which is always larger for greater flow velocity as Eq. (3.10) indicates. The paradox could be explained by the concept of discount price in *en gros* purchase.

We go to wine sales in Zurich every November to stock up our cellar. Some of the firms introduce discount prices which are inversely proportional to the dozens (cases) of bottles one purchases. For every case, a 2.5% discount is applied, or the discount price is 97.5%, until a maximum reduction to 75% of the unit price is reached if one purchases 10 cases. For still larger purchases, the same maximum discount price is applicable. Applying this concept to interpret the meaning of C_f, we might consider this coefficient a discount of unit price, the more one purchases, the less is the discount price of unit cost, until a maximum reduction is reached. After that, the same cheapest discount price is applicable. With this explanation, you can see that your total bill (F_r) is always more for more bottles purchased, even if the unit cost (C_f) is less when you purchase many cases (large **Re**).

We will return to our previous discussion when we compared the resistance coefficient to a dimensionless number used in the computation of tax rate. In that example, we recognized that the dimensionless number is not only related to the assessment rate, but also to the proportion of the taxpayer's assets in a foreign country and in the US; the coefficient for the tax assessment is larger if a taxpayer has larger foreign assets. Similarly, the resistance coefficient C_f is not only to be viewed as a coefficient of fluid friction; it is also a measure of the proportion of viscous resistance to inertial resistance; the coefficient is greater if the resisting force has a larger proportion of viscous resistance. The ratio of the viscous to the inertial force is, of course, the inverse of the Reynolds number. Therefore, the resistance coefficient is greater in flows with smaller Reynolds number, in which viscous resistance is dominant.

The linearly inverse relationship between C_f and **Re**, expressed by Eq. (3.15) can be substituted into Eqs. (3.5) and (3.6), and we have

$$F_r = 3\pi \, \eta \, D \, u \ ,$$

which is the Stokes' law of fluid resistance, or Eq. (3.3). Equating Eqs. (3.2) and (3.3), we get

$$u = \frac{1}{18} \frac{(\rho_s - \rho_f)}{\eta} g \, D \ .$$

This is the Stokes' law of settling velocity as expressed by Eq. (3.1).

After this long excursion to the realm of Newtonian physics, including a side trip to the tax office and wine ship, my students, over the years, finally have begun to appreciate the necessity of understanding a formula before they use it. Why does one obtain a ridiculously high settling velocity for sand and gravel if we use Stokes' law? The unrealistic result is the consequence of assuming a linear relationship way beyond the limit of its validity. Every customer to a wine ship knows that he cannot expect an unlimited discount at a linear reduction rate of 2.5% per case. If so, he would earn money by purchasing more than 40 cases of wine.

Stokes' law is often invoked in geology to promote the general understanding of sedimentary processes. We shall refer to this formula when we discuss the forces required to keep particles in suspension. We use the formula to calculate how much time has elapsed before tiny plankton sink to the ocean bottom after death. I would like to take this opportunity to illustrate the application of the concept to a specific sedimentological problem, namely, the reliability of varves as indicators of chronology or of past climate.

Varve comes from the Swedish word *varv* , which means a circle, and the word was introduced in 1912 by De Geer(1912) into geology to designate annually deposited thin layers of sediments. The varves are evenly laminated in lakes, and each lamina has a thickness of millimeters or less. The base of a varve is commonly a silt, which diminishes in grain size and the top is a clay, 0.004 mm in median diameter or less. De Geer reasoned that the silt was deposited in the spring and summer, having been brought to a lake by melt waters from glaciers, and that the clay was deposited during the winter. Each silt-clay pair records thus an annual circle of summer/winter deposition. Using varves, De Geer counted the number of years since the disappearance of the glacier from the Baltic Sea, and his varve chronology has been confirmed, at least in part, by radiometric dating of sediments.

The varve concept was, however, controversial. A Mr. Stumpf, working with the Swiss Hydrographic Bureau, investigated the matter by suspending a sediment trap in the 145-m deep Lake of Walenstadt in 1916. After a year, he raised his trap and examined his harvest of sediments. Laminated silt and clay were found, but he also counted five laminations. Those laminated sediments were not annual deposits, and were thus, by definition, not varves, even though they bore a superficial resemblance to varves. The observation by Stumpf was ignored because he did not have an explanation as to why he had harvested five laminated pairs of silt and clay. When Kuenen and Migliorini(1950) formulated the concept of deposition by subaqueous suspensions in 1950, they proposed that the varve-like laminated silt and clay in lakes were deposited by *turbidity currents,* or an underwater flow of suspension. Since then, the confidence of using varve chronology has been shaken. But are all varves turbidity current deposits or *turbidites*, or are there true varves?

We have been investigating the sedimentation in Swiss lakes ever since I joined the faculty here. For more than a decade, we did not find any glacial varves in our lakes. In some Swiss lakes such as Lake Zurich, there is an annual precipitation of calcium carbonate during the summer, and this process has led to the deposition of another kind of varves, the laminated lacustrine chalk; but the detrital silt/clay varves were not found. In other lakes, such as the Lake of Walenstadt, no such biochemical precipitation is taking place and no varve of any kind is present; the laminated sediments of that lake, as we have just mentioned, are not varves, but deposits from turbidity currents. My student, André Lambert (1979) and his colleagues have been able to detect such underwater currents by sinking current meters down to the lake bottom. They came a few times a year, when the Linth had spring floods, and laminated silt and clay pairs were deposited by such turbidity currents. The thickness of laminations is irregular: thick laminae were deposited by dense suspensions and stronger currents and thin laminae by dilute suspensions and weak currents. Turbidity currents have also been monitored in other Swiss lakes; they occur at times of high flood discharge (Fig. 3.5).

We encountered for the first time in 1980 sediments which truly looked like varves of De Geer (Fig. 3.6a). We then drilled a hole through the bottom sediments at a spot where Lake Zurich is deepest (145 m). The hole penetrated a sedimentary sequence deposited during and since the last glaciation. This varve-like sequence was deposited some 13000 to 11000 years ago, shortly after the Linth glacier retreated from the lake basin. The Lake Zurich laminations are always topped by very thin laminae of clay. In this respect they differ from those sampled from the Lake of Walenstadt, where the top of the laminae could be a fine silt or even a coarser silt.

I was not convinced that we had varves, but an academic guest from China, Zhao Xiafei, convinced me that they are truly annual deposits. He pointed out that the varves are not characterized by their summer deposit, when one or more layers of silty sediments could be deposited; varves should be distinguished by their winter deposit when a lamina of clay, with a median diameter of some 0.002 mm, is deposited. Zhao made grain analyses of the clays from the Zurich laminated sequence. He found indeed such winter deposits; no other clays from other Swiss lakes have a median size as fine as that of the varve clay.

"Why?" I asked him.

"I don't know; the observation is empirical."

After thinking over the question, the answer is very simple. Lake Zurich is situated at an elevation of about 400 m above sea level, and the northern European climate is such that this deep lake does not freeze in winter, except once every 30 or 40 years. The main seasonal difference in the sedimentary condition is the chemical saturation every summer which produces the summar lamination of lacustrine chalk. The other lakes investigated by us, such as the Lakes of Walenstadt, Thun, Brienz, Lugano, Geneva, Konstanz, etc., were also situated at a low elevation of several hundred meters, none of

Fig. 3.5. Rhone River discharge and turbidity current in Lake Geneva (after Lambert and Giovanoli 1988). Current meters detect current events in Lake Geneva at times of high Rhone River discharge, and the maximum speed is almost 1 m/s. The *thick bars* indicate the speed recorded by the upper current meter and the *thin bars* indicate the speed of the lower. Since the upper meter was anchored 13 m above the lake bottom, the data indicate that the larger currents were at least that thick

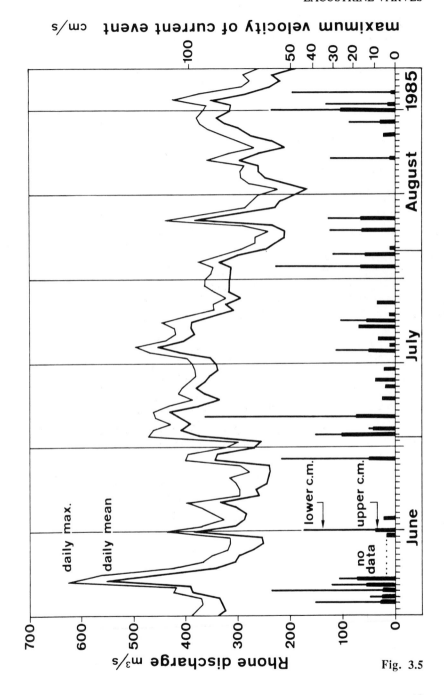

Fig. 3.5

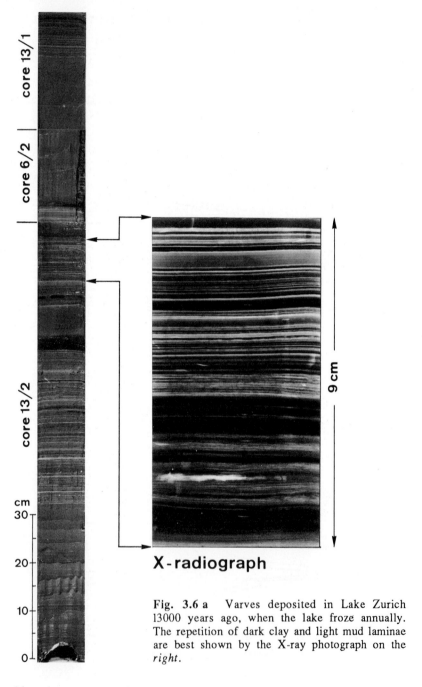

X-radiograph

Fig. 3.6 a   Varves deposited in Lake Zurich 13000 years ago, when the lake froze annually. The repetition of dark clay and light mud laminae are best shown by the X-ray photograph on the *right*.

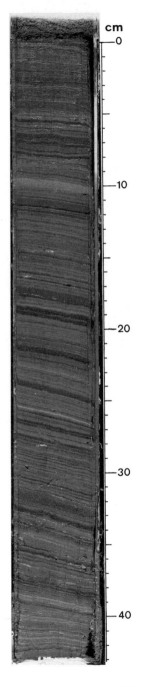

Fig. 3.6b Varves deposited in a high-altitude Alpine lake (Lake Oeschinen, Kt. Bern, 1578 m above sea level) in modern times when the lake freezes annually

those freeze regularly every winter. So we have never found varves in those lakes which do not freeze every winter.

Studies of the Lake Zurich sediments have permitted us to reconstruct the climatic history of the region. Some 17000 or 16000 years ago, the bottom of a valley glacier melted so that a subglacial lake was formed. With the retreat of the glacier, the lake basin was covered by stagnant ice. Warming led to occasional thaw of the frozen lake, perhaps once every 30 or 40 years. Some 13000 years ago, the climate of the Zurich region should have been similar to the present climate of the Engadine Valley of Switzerland, where the deep lakes freeze every winter, and thaw every summer. This situation lasted a few millenia or so, before the continued warming led to the climatic conditions of today; that was the beginning of the *Holocene*. Since then, the lake seldom freezes except occasionally at irregular intervals. The correlation of sediment and climate indicates that the Lake Zurich varves were deposited at a time when the lake froze every winter.

Stokes' law provides the theoretical explanation of the origin of varves. When the lake is not frozen, sands, silts, and clays are all brought in by rivers, and the sediment deposited from surface suspension includes silt- and clay-sized particles. After the lake is frozen, no further sediment is brought in until the thaw 3 or 4 months later. During the winter season, sedimentary particles larger than silt size (0.004 mm) should all have settled down, within a month after the lake freezes, to the bottom of this 145-m deep lake. The rest of the suspension which is sedimented during the months after the freeze should produce a sediment with a median grain diameter of 0.002 mm or less. That is the winter deposit of a lake which freezes in the winter. In a lake which freezes every winter, the winter deposit marks the top of each varve.

Since then, we have checked the prediction with new observations. We found that the varves of Lake Lugano were deposited some 13000 to 11000 years ago, when that lake, like Lake Zurich at that time, also must have frozen annually.

Scientific theories predict. We never found varves in modern sediments because we had been studying Swiss lakes at lower elevations; those lakes rarely freeze. The climate in mountain country above 1600 m elevation is such that the lakes there freeze every winter. If the theory is correct, varves should be found there, if the lake water is deep enough to contain enough suspension to deposit a clay lamina during the freeze. We started a campaign to core the sediments of Silvaplana Lake, St Moritz Lake, the lake on Bernina Pass, and other deep lakes at high altitude. As we predicted, we found varves in each of those lakes, which are detrital varves like those deposited in Lake Zurich or Lake Lugano 13000 years ago (Fig. 3.6b).

We now believe that varves are annual deposits. Furthermore, true varves are a wonderful climatic indicator. The climate of varve deposition should be similar to that in the Engadine Valley of Switzerland today, where the lakes freeze every winter. In colder regions like the Arctic where lakes rarely thaw or in warmer regions like the Swiss Midland where lakes rarely freeze, no varves are deposited.

Suggested Reading

This chapter is the essence of this book. I have written and revised the text almost a dozen times over the last two decades. A student is advised to go over the text repeatedly before they delve into other publications.

For students who do not appreciate the difference in scientific approach, they can look up other discussions of the Stokes' law, such as that by M. R. Leeder in *Sedimentology* (London: Allen and Unwin, 1982, p 75), by J. R. L. Allen in *Physical processes of sedimentation* (London: Allen and Unwin, 1970, p 46), or by Gerald Middleton and John Southard in their *Mechanics of sediment movement* (Binghamton, NY: Eastern Section, SEPM, 1978).

I have touched upon the subject of dimensional analysis, but I cannot introduce the essence of the methodology, which is based upon an assumption that certain variables are the independent variables of a phenomenon, and that all others are dependent variables, redundant or irrelevant. For those who wish to delve deeper into this mathematical approach to sedimentological problems, I recommend a book by H. L. Langhaar on *Dimensional analysis and theory of models* (New York: Wiley, 1951).

The basic principles of hydrodynamics are discussed in detail by L. Prandtl, K. Oswatisch, and K. Wieghardt in their *Strömungslehre* (Braunschweig: Vieweg, 1969); there must be an English translation of that famous textbook.

For advanced students interested in the sedimentology of varves, they should refer to the collection of papers in *Moraine and varves* (Ch. Schluchter(ed), Rotterdam: Balkema, 1979) , and an article by X. F. Zhao, K. J. Hsü, and K. Kelts on *Varves and other laminated sediments of Zübo* (Stuttgart: Schweizerbart'sche Verlag, Contributions to sedimentology 13: 161-176, 1984). The investigations of riverborne turbidity currents in Swiss lakes are summarized by a Lambert and Giovanoli paper in *Limnology and oceanography* (33:.458-468, 1988).

4 Sediments Are Moved

Turbidity Currents - Chezy's Equation - Stream Transport
Shield's Diagram - Hjulström's Curve

The science of sedimentology was revolutionized in 1948, when the concept of turbidity-current transport and deposition was introduced by Kuenen and Migliorini(1950). Turbidity currents, which are not observable in nature, are supposedly generated by submarine slides of catastrophic proportion. Such a postulate was a radical departure from the uniformitarianism of geology preached by Charles Lyell, who stated that processes operating in the past are those observable in the present, with the same energy or intensity. Therefore, the new concept was controversial in the 1950s, when I was a young sedimentologist working for the oil industry. The controversy was not only aired at professional meetings, but also during coffee breaks and cocktail hours.

"What are turbidity currents?" " How big are they?" "How fast do they move?" Those were the questions which were often asked.

Turbidity currents are underwater floods. A river flood entering a lake with a suspension which has a density greater than that of lake water will, for example, flow along the lake bottom until its load is deposited. We lose sight of the turbidity current when it plunges into the deep, and discovery of this unobservable sedimentary agent is an interesting chapter on the history of science, as described in a 1985 review article which I co-authored with Kerry Kelts. Observing that the meltwater torrent of the Rhone disappears under the clear water of Lake Geneva, August Forel proposed in 1885 that the silt-laden stream must have continued downward as a suspension current. Such a current could cut a subaqueous channel on the lake bottom and deposit its load in the lake. This suggestion had been generally ignored until half a century later, when R. A. Daly resurrected Forel's hypothesis and applied the idea to explain the origin of submarine canyons. Another decade and half had gone by before the importance of a current as a depositing agent in deep seas was recognized by sedimentologists studying ancient and recent marine sediments.

The presence of turbidity currents in the Lake of Walenstadt, as mentioned in Chapter III, has been confirmed by current-meter measurements (Lambert 1979). We made our first measurements in 1973, when the Linth River of Canton Glarus, which empties into the Lake of Walenstadt, reached a flood stage after several rainy days during the snow melting season. Our record, from a current meter anchored on the lake bottom at a site some 4.5 km distant from the mouth of the Linth River, showed that an underwater current, or *underflow*, was active. This underflow on the lake bottom was a few meters thick and reached a maximum speed of 30 cm/s. A second survey was carried out, again after the Linth reached a flood stage, during late July of 1977. The record registered no current movement until 15:30, July 31, when a current, with a speed ranging from 20 to 50 cm/s, was detected at the survey station. A comparison of this current record with the gauge record of the river discharge suggested that the turbidity current must have been the subaqueous continuation of the peak flood of the Linth (Fig. 4.1). The underflow arrived at the lake-bottom station 1.5 hours after travelling 4.5 km from the Linth measuring station. The average speed of the current front was thus 28 cm/s, within the range of the measured speed. A layer of silty mud was deposited by this current, which, as we have discussed, resembles a varve but is not a varve.

That river water carrying suspended particles emptying into a water lake should sink and become an underwater river is easily understood, the suspension has a higher density than 1.0 g/cm^3. However, riverwaters seldom carry enough suspension to possess a density greater than normal seawater. When the Mississippi River flood reached the Gulf of Mexico, for example, the floodwater did not sink, but spread out to form a *plume*, which is a jet-like freshwater layer above the denser seawater. Floodwaters rarely, if ever, sink to form turbidity underflows in oceans. The origin of marine turbidity currents is thus commonly related to submarine sliding.

In 1929 trans-Atlantic telegraph cables were broken after a large earthquake shocked the Grand Banks area. They were broken successively from north to south (Fig. 4.2). The event was a puzzle and seemed to defy an explanation. More than 20 years later, Philip Kuenen(1952), who had championed the importance of turbidity currents as a sedimentary agent, discussed this with Maurice Ewing and Bruce Heezen of the Lamont Geological Laboratory.

Fig. 4.1. Turbidity current in the Lake of Walenstadt is a continuation of river flood. The *upper diagram* shows the discharge of the Linth River due to heavy rains of July 31, 1977. Note that the flood peaks shortly after midday. The underflow is registered by the *lower diagram*, which shows that it arrived at the site of the current meter, at about 2 p.m., 1.5 h after traveling 4.5 km from the Linth measuring station

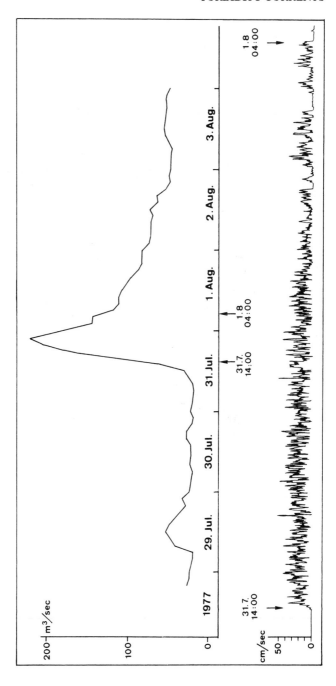

Fig. 4.1

Kuenen(1952) convinced the Lamont scientists that the cable breaks were the work of a fast-moving turbidity current. They postulated in 1952 that the earthquake triggered a large submarine sliding movement, which in turn generated a turbidity current. Using the data on the timing of the moment when telegraphical messages could no longer be sent across the Atlantic, the speed of the turbidity-current flow was computed. It was found that the speed of the moving mass at 55 knots at the foot of the steep continental slope, and was still 14 knots, or almost 30 km/h, after some 500 km of transport (Fig. 4.2).

Now that estimates on speed were obtained, Kuenen proceeded to cite a modified version of Chezy's equation in an attempt to estimate the size and density of the current.

Chezy's equation states

$$u = C \sqrt{d\,s} \ , \tag{4.1}$$

where u is the current velocity, C Chezy's coefficient, d the depth of an open channel, and s the slope of the channel floor.

Realizing that the effective weight of an object in water is $\Delta\rho = (\rho_c - \rho_f)$, Chezy's equation has been modified so that

$$u = C \sqrt{\Delta\rho_c\,d\,s} \ , \tag{4.2}$$

where u is the flow velocity of turbidity current, $\Delta\rho_c$ the effective density, and d the thickness of the current. Making some assumptions, Kuenen arrived at the startling conclusion that the Grand Banks current could be several hundred meters thick, and up to 1.6 g/cm^3 in density.

I came across Chezy's equation for the first time when I read Kuenen's 1952 article. At the time I had completed my Ph.D. studies and was employed as a research sedimentologist. I had, however, no idea where the equation came from.

Fig. 4.2. The Grand Banks earthquake of 1929 and subsequent cable breaks. Longitudinal profile of the ocean bottom along the postulated paths of the sediment-gravity flow and turbidity current. *Arrows* show the position or projected position of the breaks. The timing of those not broken immediately are indicated (e.g., 00:59, signifying 59 min after the earthquake). The ratio (e.g., 1:20) shows the slope gradient. Gravel deposit (G) is present at the base of the slope, and turbidite (T) on the abyssal plain. The postulated speed of the current (in knots) is shown by the upper diagram. The profile has a vertical exaggeration of 60x

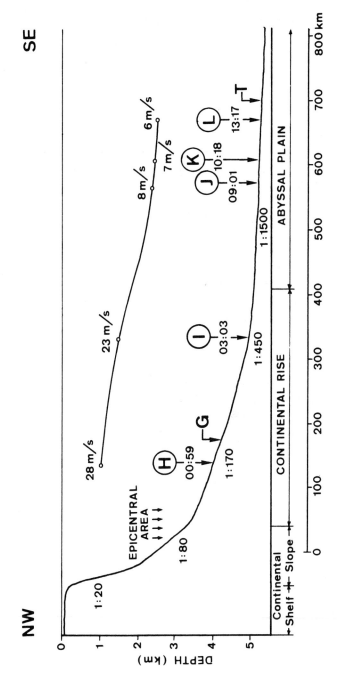

Fig. 4.2

There are four variables, u, $\Delta\rho_c$, d, s, and one numerical coefficient in the equation. Of the four variables two (u, s) are known and two ($\Delta\rho_c$, d) are unknown. There cannot be a unique solution of an equation with two unknowns. In order to obtain a reasonable d value of 100 m or less, Kuenen had to assume a density of 1.6 g/cm^3 for the turbidity current at the more distant cable breaks. Such a high density was incredulous, if not impossible.

When unreasonable numerical results are obtained from the application of a formula in making calculations, everyone's first reaction is to examine its applicability. Is Chezy's equation applicable? I began checking Kuenen's calculations by laboratory experiments, and a flume was built to observe artificially produced turbidity currents.

A sediment-water mixture with a density of 1.6 g/cm^3 is a mud, and it simply does not flow as a suspension. To obtain a good suspension current in my flume, I had to use dilute mixtures of mud and water, with a density of about 1.05 g/cm^3 or less. Later, I learned from the results of experiments carried out by R. A. Bagnold(1962) that sediment with a volume concentration greater than 9%, or a mixture with a density greater than 1.1g/cm^3, would, as a rule, not move in suspension. Concentrated dispersions of grains tend to move by grain collisions near the channel bottom, and this mode of motion is now called *grain flow*.

Even for dilute turbidity currents, I cannot use Chezy's equation to relate the various known variables in my experiments. Equation (4.2) states that the speed of the flow should be zero when the bottom of the flume is horizontal. Yet the experimentally produced turbidity currents will flow at a finite speed when the flume floor is flat. In fact, experimental turbidity current flows under its own momentum upslope. Obviously, I had to find out more about Chezy's equation, before I could understand the discrepancy between my experimental observations and Kuenen's calculations.

I checked my old lecture notes and I consulted textbooks on sedimentary rocks, but I could not find Chezy's name. Someone told me to look through a textbook on hydraulic engineering. I found the equation and understood its derivation from a Newtonian consideration of equilibrium of forces.

Concerned with the water supply to Paris, the administration entrusted Antoine Chezy in 1768, an obscure school-teacher, the task of designing a canal to the city. The channel had to have a sufficiently large cross-section, and a sufficient gradient for enough water to flow through it to supply the city's need. Yet, the slope must not be too steep, otherwise the acceleration would lead to overflow and flooding of the countryside.

Chezy could not find any ready made formula in the engineering handbooks of his days. The idea occurred to him, however, that a solution may lie in a suitable comparison. Applying Newton's First Law, he recognized that whatever the initial velocity may be, it diminishes or increases rapidly enough to become a uniform and constant velocity when the gravity is balanced by fluid resistance. For the water in a channel segment of unit length, the moving force is the downslope component of the weight, which is the mass of the moving water, ρV , times the gravitational acceleration g,

$$F_g = \rho \ V \ g \ \sin \theta = \rho \ d \ w \ L \ g \sin \theta \ , \qquad (4.3)$$

where V, the water volume flowing through a rectangular channel, is the product of the width w , the depth d , and the length L.

The resistance in counterbalance is, as we have explained in the last chapter (Eq. 3.7),

$$F = C_f \frac{\rho u^2}{2} \ A \ ,$$

where C_f is a dimensionless resistance coefficient, apparently comparable to, but not quite the same as the coefficient of friction of solid sliding on solid, and A is the area of the channel in contact with that volume of moving water. This area is equal to the unit length times the *wetted perimeter*, R, which in the case of a rectangular channel is (see Fig. 4.3):

$$R = w + 2d \ .$$

Equating driving force and resistance, we have

$$\rho \ d \ w \ L \ g \sin\theta = C_f \frac{\rho u^2}{2} (w + 2d) \ L$$

or

$$u^2 = \frac{2g}{C_f} \cdot \frac{d \ w}{(w + 2d)} \cdot \sin\theta \ .$$

The ratio $d \ w \ / (w+2d)$ has also been called the *hydraulic radius* of a channel. In the case of a channel with a depth much smaller than its width, or for very large w, the value of the hydraulic radius should be about the same as d , because w of such a channel is about the same as $(w + 2d)$. In other words, for the sake of rough calculations, the value of depth d could be substituted for that of the hydraulic radius of wide and shallow channels. For a very gentle incline, its slope $\tan \theta$ is approximately equal to the sinus of

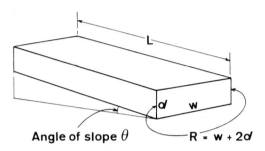

Angle of slope θ R = w + 2d

Fig. 4.3. Geometry of an open channel and definition of a wetted perimeter. The length of a channel segment is represented by **L**, the channel width by **w** and the channel depth by **d**. The wetted perimeter is the circumference of the channel below the water, and, in the case of a rectangular channel, is (w + 2d). The wetted surface area of the channel segment is (w + 2d) L. The cross-sectional area of the channel is w · d

angle θ. Taking into account both approximations, Chezy's formula can thus be written:

$$u = \sqrt{\frac{2g}{C_f}}\sqrt{d\,s} \qquad .$$

(4.4)

Introducing a new parameter C in place of $\sqrt{2g/C_f}$, and calling it the *Chezy coefficient*, we have Chezy's equation, as shown by Eq. (4.1):

$$u = C\sqrt{d\,s} .$$

This equation expresses the steady-state velocity of open-channel flow. In applying this relation to turbidity currents, Kuenen took note of the fact that the moving force is $\Delta\rho_c Vg\sin\theta$, and that $\Delta\rho_c/\rho_c$ is approximately $\Delta\rho_c$ for dilute currents.

What are the numerical values of Chezy's coefficient? Kuenen wrote that the value of "C for large rivers is 700-800". He did not cite the source of his information. I looked up in a textbook, which is written for American engineers, and found a quite different range of C values for large rivers. I puzzled over the problem before I realized that the Chezy coefficient is not dimensionless. The value of C is related to the unit of gravitational acceleration, which is 981 cm/s^2, but it is 32 f/s^2. The C values cited by Kuenen are expressed in square root of centimeters per second and those by the American engineers are expressed in the square root of feet per second.

Why should the values of C for large rivers be 700-800 cm$^{-1/2}$/s? This is determined on the basis of field observations of the speed related to the depth and slope of the channel. Going back to the definition of the Chezy's coefficient

$$C = \sqrt{2 g / C_f} \; , \qquad\qquad (4.5)$$

C values of 700-800 cm/s correspond to C_f values of 0.003-0.005. Field measurements have thus more or less verified the experimental results.

A dimensionless coefficient, called friction factor f, for pipe flow has been obtained experimentally by various hydraulic engineers, Nikuradse in Germany and Moody in the United States (Fig. 4.4). Curves relating the coefficient to the Reynolds number of fluid flows are known to American engineers as *Moody diagrams*. As will be shown later, the resistance coefficient of open-channel flow is related to the friction factor by a factor $1/4$. The C_f values for river flows are estimated on the basis of such experimental data on the friction factor.

Chezy's equation has an apparent elegance of simplicity. In fact, the substitution of a dimensionless resistance coefficient by a coefficient with an irrational physical dimension of $L^{-1/2}t^{-1}$ causes confusion. Nevertheless, an equation relating stream velocity to its depth and slope serves readily to explain some alluvial processes.

As you may have learned from geography lessons, the two main types of streams are meandering streams and braided streams. Why do streams meander and why do they become braided?

Meandering has caused many political and legal problems, because many streams serve as international or property boundaries. The mechanics of meandering has been a favorite study for hydraulic engineers. For us, it is sufficient to learn that meandering is a need for a stream to find equilibrium again.

A stream channel, like Chezy's canal for the water supply of Paris, has a depth and slope designed to permit the flow of a certain volume of water at a certain speed. At times of flood, the volume of water increases. The moving force becomes greater, the current velocity has to increase. Chezy's equation tells us that the increase could be accomodated by increasing either the water depth or the slope of a streambed. In actuality, the latter cannot be done

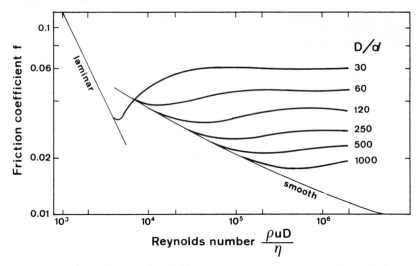

Fig. 4.4. Variation of the friction factor **f**. The values of the friction factor to fluid flow in open channels have been determined experimentally. This diagram illustrates the results of J. Nikuradse in 1932 and 1933. The ratio D/*d* (pipe diameter / roughness) gives a measure of the roughness of the inner surface of the pipe

overnight, if at all. The consequence of flooding is, therefore, always an increase of water depth. So the water overflows the river banks.

Flood catastrophes come when the flooding was sudden. Otherwise, streams find their equilibrium, not by a depth increase, but a slope change. The stream can adjust to its increasing velocity by cutting a steeper streambed, or the torrent of flood can be slowed down if the stream course is modified to give a more gentle gradient. To cut a ravine is a common phenomenon in the mountains; we see the deep erosion by floods after deforestation of steep slopes in our mountains. In the plains country, streams do not cut down; they move sideways, or they meander, to reduce the gradient and thus the speed of flow. Meandering serves the same function as slalom skiing to reduce the speed of descent, and the sideward erosion is readily accomplished in places where the riverbanks are underlain by unconsolidated river sediments. Cave-in at times of flood is a manifestation of this process. When I was a child in China, I heard stories of an entire village on the banks of the Yellow River disappearing into the floods. It is thus not surprising that the Yellow River delta is deserted, since no one can be sure whether the fruit of his labor would not be washed away by the next flood.

Civilization interferes with natural processes. The Rhine channels at many places in Switzerland have been cemented to prevent the erosion of its riverbanks. The stream cannot meander to smother the fury of a flood and it is piped down until it inundates the flat land. I was on vacation in Cologne in 1988 and was told that the Rhine had reached a historically high level. We did have a heavy rainfall in the spring, but there should have been even heavier spring precipitation during the 2-millenia history of Cologne. Excess precipitation alone cannot cause the record high of the river level in Rhineland; part of the problem at least could be traced to the fact that riverbanks are cemented upstream. When flow speed cannot be reduced by meandering, the excess volume of water rushes downstream to cause flooding.

Braided streams are commonly found in more arid countries where the riverbeds are broad and flat. Normally, waters trickling out of springs gather their way into one small shallow stream. I remember my student days working in the San Antonio Canyon of the San Gabriel Mountains in California. The valley is almost 1 km wide, but I could almost jump across the tiny San Antonio Creek. Once after a heavy rainfall, I went into the canyon again. I found not one but a dozen streams in the valley. The floodwater had not been totally channelized into the normal course of the creek to increase its depth, but was distributed, having found its way into the various braided channels. Consequently, neither the depth (d), the slope (s), nor the speed (u) of the San Antonio Creek itself increased very much, but the total volume transport through the various braided channels was at least an order of magnitude more than that transported normally.

Sediments are transported by streams, and such sediment transport is an alluvial process. Streams vary in velocity, kinetic energy, and power. The size and quantity of sediment particles transported are related to those physical variables, and two terms have been introduced: *competence* and *capacity* .

These words are used in everyday life, often in connection with financial transactions. When I am given by my university an ordinary budget of 25,000 Swiss francs a year, that is the *capacity*; I cannot spend more than that amount annually for my ordinary teaching and research activities. I also cannot spend all that money on one big check; I have only the *competence* to sign the bill for an item costing less than 5000 Swiss francs from this budget. For more expensive items, I have to make request for extraordinary funding. In other words, the total amount is limited by the capacity and the largest single check by the competence.

We can illustrate the principle by material transport. Suppose you would like to build a vacation house in the mountains, and the building materials are to be transported first by trucks and then by a fleet of tractors. You ordered building stones of various sizes as well as sacks of sand and cement. When the materials carried by the trucks reached their destination at the valley station, you found to your dismay that some very big stones and many sacks of cement have to be left behind; the stones are too heavy, and the sacks are too many. None of the tractors have the competence to carry stones larger than a certain size, and the number of tractors too few, or their transport capacity insufficient, to transport all the cement sacks. The sediment transport by rivers is governed by the same principle. Boulders are left behind in the mountains, because they are too heavy to be moved; their transport has been limited by stream competence. Not uncommonly, however, mountain streams have a sandy bottom; sands have been left stranded because of the limited transport capacity.

I tried out this lecturing approach on my son Peter, who was finishing middle school. He said that the limitation of cement transport was only temporary. When the tractors came back the second time, they could take away the sacks which had been left behind. He is correct, the accumulative capacity is time dependent: Sand deposited on a river bar at one flood, may be carried away by the next flood. To move giant boulders is, however, not a question of capacity, but of competence. Blocks of landslide debris, fallen down from cliffs, are commonly too huge to be moved by stream power, they stay where they are before they are weathered away, or until they are buried by younger sediments.

What force is required to set sedimentary particles in motion? Both hydraulic engineers and sedimentologists have done experiments to investigate this problem. We learn in textbooks of Shields' criterion that the critical current velocity to move a sediment particle and the particle size are related by the equation

$$u^* = 0.06\sqrt{(\rho_s - \rho_f) g \cdot D} \quad , \tag{4.6}$$

where u^* is called shear velocity, D is the diameter of a spherical particle, and 0.06 is apparently an experimental constant. How did Shields come to this conclusion?

When a grain begins to move, its inertia has to be overcome, and this resisting force is

F_r = dimensionless number · effective weight of the grain

or

$$F_r = Z (\rho_s - \rho_f) g D^3 \quad , \tag{4.7}$$

where D is the diameter of the grain. The resisting stress is force per unit area, or

$$\tau_c = C_s (\rho_s - \rho_f) g D \quad , \tag{4.8}$$

where C_s is a dimensionless number.

The magnitude of shear stress exerted by a fluid could be experimentally measured, but it is difficult for a person to think in terms of stress, be it in dynes/cm^2, or Newtons per square meter. It is easier, however, for a person to appreciate the significance of numerical expressions for velocity; we can easily visualize cm/s or m/s or km/h. Instead of talking about the force or stress needed to move a sedimentary particle, we prefer to relate that to the velocity of fluid flow.

The fluid stress and velocity is related by Eq. (3.7)

$$F_i = \tau\, A = C_f\, \frac{\rho\, u^2}{2}\, D^2 \quad,$$

where τ is the applied stress and A is the cross-sectional area of the particle. When the shearing stress exerted on the grain by the fluid force reaches a critical magnitude, τ_c , the grain should begin to move. The stress shearing the grain has the same dimension as the fluid stress, so that we have

$$\tau_c = \text{dimensionless number} \cdot \frac{F_i}{A} = Z\, \frac{\rho\, u^2}{2} \quad, \qquad (4.9)$$

where Z is a symbol for a dimensionless number, of an indefinite value. Expressed in terms of fluid velocity, we have

$$u_c = Z\, \sqrt{\frac{\tau_c}{\rho}} \quad. \qquad (4.10)$$

Engineers introduced the concept of *shear velocity* , u^*, to designate the critical velocity needed to initiate grain motion. This conceptual velocity is defined by relating

$$u^* = \sqrt{\text{critical shear stress/ density of fluid}} \quad. \qquad (4.11)$$

Substituting Eqs. (4.8) and (4.10) into this definition, we have

$$u^* = \sqrt{C_s\, \frac{(\rho_s - \rho)}{\rho_f}\, g\, D} \quad. \qquad (4.12)$$

For a critical shear stress of 10 N/m^2 or 100 dynes/cm^2, the shear velocity is 0.1 m/s or 10 cm/s.

The dimensionless number C_s in Eq. (4.8) is the ratio of the critical stress τ_c, and the parameter $(\rho_s - \rho_f)g\, D$. The value of this number can be

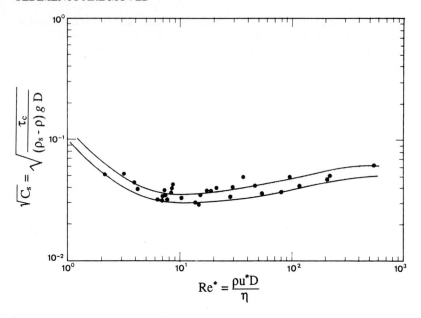

Fig. 4.5. Shields diagram (modified after Shields 1936). Shields found that the relation between the shearing stress required to move a particle and its diameter could be expressed by a parameter which is a dimensionless number C_s. Experimenting with grains of various density, such as amber, lignite, graphite, barite, as well as quartz sand, Shields obtained an empirical relation between this coefficient and Reynolds number. The value of C_s is about 0.06 for stream flows, which have large **Re** values

calculated on the basis of measured τ_c for the grains of given ρ_S and D.

The value of C_S varies according to the flow characteristic near the boundary to a stream bed. The concept *boundary Reynolds number* is thus introduced by substituting shear velocity u* for the average velocity of the fluid flow [Eq.(3.13)]. We have thus:

$$\mathbf{Re}^* = \rho \, u^* \, D / \eta \; . \qquad (4.13)$$

Experiments by Shields revealed that the values of C_S fall commonly within a range between 0.03 and 0.1 depending somewhat upon the value of the boundary Reynolds number; the experimental results are shown by the so-called Shields' diagram (Fig. 4.5). The diagram shows, furthermore, that for turbulent flows with $\mathbf{Re}^*$ values larger than 1000, the value of C_s is fairly constant; its square root has a value of about 0.06. Substitute this numerical result of experimentation into Eq. (4.12), we then have Eq. (4.6) stating the critical shear velocity necessary to move a sedimentary particle of a certain size.

What is the relation of this theoretical shear velocity, u^*, to the actual velocity of stream flow? A comparison of Eqs. (4.10) and (4.11) indicates $u = Z u^*$ or the fluid velocity is a multiple or a fraction of u^*. We know that the linear velocity of the turbulent water of the stream is varied. The velocity increases from the bottom to a maximum at some depth toward the surface of the stream (Fig. 4.6). We could obtain an average linear velocity of a stream on the basis of the rate of volume transport per cross section. Comparing this average velocity to the calculated value of shear velocity, Middleton and Southard found in 1977 that the average velocity of the experimental currents are several times larger than that of the u^* in their experiments. We should, therefore, not lose sight of the fact that the critical shear velocity is not the velocity measured directly by observing the distance of flow per unit time. This shear velocity is calculated from stress measurements made by transducers placed at the sediment-water interface in flumes. The average flow velocity may have been several times larger than the calculated critical shear velocity necessary to move a spherical particle of diameter D.

Experiments relating the threshold shear stress and grain diameter of sediment moved by a current have been carried out. The experimental data form the basis of using sediment grain size as the "paleocurrent meter" to estimate the average speed of ancient currents (Fig. 4.7).
Earth scientists have also conducted experiments to directly relate the velocity of fluid flow to the grain size of bottom sediment. From Eq. (4.6),

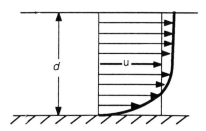

Fig. 4.6. Velocity profile of open-channel flow. The average velocity is the rate of volume transport per cross-sectional area, and it can be calculated by the relation that the area u d in this figure is equal to the area under the actual velocity profile

we expect that the flow velocity necessary to erode the bottom of unconsolidated sediment should correlate to its grain size. The results obtained by a Swedish geographer, Hjulström, are shown by Fig. 4.8, which has been called Hjulström's curve. The trend predicted on the basis of Shields' diagram holds for sand grains and gravels, but not for silt and clay particles.

This deviation from the prediction is easily explained by the fact that cohesion between clay grains is the force which we have not considered in our consideration of the balance of force. Finer sediments, such as clay-sized

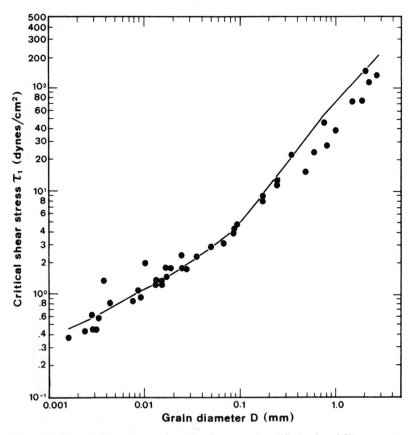

Fig. 4.7. Threshold stress and grain diameter (modified after Miller et al. 1977). Laboratory experiments, with quartz grains in water of 20°C, yielded data relating critical shear stress to grain diameter. The numerical values of the critical shear velocity (in cm/s) are related to the square root of the values of the critical shear stress shown by this diagram

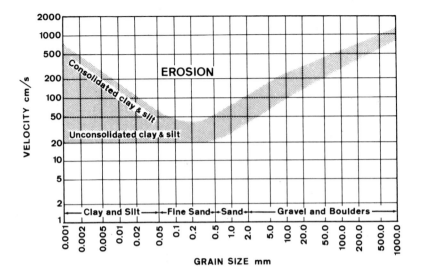

Fig. 4.8. Hjulström's curve (modified after Sundborg 1956). Hjulström determined the critical velocity necessary for moving quartz grains at 1 m water depth in open channels; the *shaded area* indicates the scatter of experimental data. This graph shows (1) that clay and silt particles are more difficult to erode than predicted, and (2) the average current velocity to move a grain is several times the calculated critical shear velocity

particles, are more difficult to erode than fine sand grains, especially when they are somewhat consolidated, because a cohesive force exists between the clay flakes. Sticky mud resists wave erosion more effectively than sand, which I noted, for example, when I studied the Recent Gulf Coast sediments. Marshes in numerous parts of western Louisiana are not fringed by sand beaches, because sand grains have been eroded away while the mud has remained.

Suggested Reading

We have discussed in this chapter several basic principles of hydraulics. Advanced students may wish to look up the *History of hydraulics* by H. Rouse and S. Ince (Paris: Blanche, 1956). I always prefer the historical approach, because a student can better appreciate the fact that physical laws are not revelations of God. They have been formulated by people for specific natural phenomena with restricted ranges of validity. An excellent book on the application of hydraulics to geologic problems is, in my opinion, *Fluvial processes in geomorphology* by Luna Leopold, Gordon Wolman and John Miller (San Francisco: Freeman, 1964)

Shields' diagram was published in *Anwendung der Aehnlichkeitsmechanik und der Turbulenzforschung auf die Geschiebebewegung* (Mitt Preuss Versuchsanst Wasserbau Schiffbau, Berlin, Vol 26, 1936); the report was translated into English and issued as Report No. 167 of the Keck Laboratory of Hydraulics, California Institute of Technology, Pasadena, California. The Nikuradse curves were published in *Strömungsgesetze in rauhen Rohren* (Verein Dtsch Ing, Forschungsheft 361, 1933), and the Hjulström diagram in an article on *Transportation of detritus by moving water* (Trask volume on *Recent Marine Sediments*, 1939). An excellent, up-to-date summary of the *Threshold of sediment motion under unidirectional currents* was authored by M. C. Miller, Nick McCave and Paul Komar (*Sedimentology*, 24: 507-527, 1977).

The classic paper by Ph. Kuenen *on the Estimated size of the Grand Banks turbidity current* appeared in the American Journal of Science (250:874-884, 1952). Our work on lacustrine turbidity currents is summarized in an article by Hsü and Kelts entitled *Swiss Lakes as a geological laboratory* (Naturwissenschaft 72: 315-321, 1985).

5   Rocks Fall

Sediment-Gravity Flows - Elm Landslide - Grand Banks Slide
Speed of Slides - Debris Flows - Sand Avalanches
Mud Slide - Olistostrome

When you see a gravel in an alluvial conglomerate, you might try to estimate the speed of river currents on the basis of the Shields' diagram, but can you use the same approach to estimate the speed of *sediment-gravity flow* on the basis of the largest clast in a breccia formed by such a mechanism? The answer is no.

What is a sediment-gravity flow? Almost all flows are produced by the effect of gravity, either directly or indirectly. In the case of fluid flows, or fluid-gravity flows, the gravity acting on the fluid moves the fluid. We have seen that the moving force of a stream is a component of the weight of the moving water.

In the case of sediment-gravity flows, the gravity acting on the sediment, or the weight of particles, angular fragments, boulders, pebbles, sand, or dust, moves the flowing mass of solid particles. This type of movement is familiar to skiers who are often witnesses of snow avalanches. An avalanche moves when a mass is separated from a cliff, falls down, and is broken into countless fragments or sediment particles of various size. The gravitational potential energy of the mass is changed into the kinetic energy of the motion, and the changing momentum of the motion is the force which compels the particles to move.

Rock-avalanche deposits, with debris flowing hundreds of kilometers, have been found on the Moon and Mars. The debris were carried neither by air nor by water; there is neither on those heavenly bodies. On Earth, an avalanche mass would soon be mixed with an ambient fluid, be it water or air. The movement of a solid-water mixture tumbling down a steep slope may, however, not be a liquid flow, because it is not necessarily the gravity acting on the water which makes the solid particles move. In very dense mixtures, it is the gravity acting on the sediment which makes the water move. Similarly, in the case of a solid-air mixture, such as a snow avalanche, it is the snow that moves air; snow is not carried by the air.

When I was a young student working in the desert of western North America, one of our favorite pastimes was boulder-rolling. I remember particularly a steeply dipping bedding surface, exhumed by erosion, on the side of the Wasatch Mountains in Utah. We should have been doing field mapping, but often we engaged ourselves in a contest to see whose boulder would go the farthest. Instinctively we all tried to find the heaviest boulder, but the biggest was not always the winner. In the downslope movement, the main resistance force is friction while air resistance is negligible; a well rounded cobble may have the least rolling friction.

The physics of motion down an inclined plane is usually taught in middle school: At the moment when a sliding block begins to move, under its own weight, down an incline

$$F_g = m \ g \ \sin\theta_c \tag{5.1}$$

$$F_r = \mu \ m \ g \ \cos\theta_c , \tag{5.2}$$

where F_g is the gravity, F_r is the resistance, μ is the coefficient of friction, and θ_c the angle of repose. Equating the two forces, we have

$$\mu = \tan \theta_c , \tag{5.3}$$

namely, the tangent of the critical *angle of repose* is the coefficient of friction.

For boulders or sand grains sliding down the lee face of a dune, the angle is a slightly over 30°, and the coefficient of friction is thus about 0.6. This angle is smaller if the sliding surface is wet; it could be 20° or less.

Land slides or rock falls when they lose cohesion along a steeply dipping fracture plane. The rock mass becomes detached, breaks up, and "drives" its way down the slope. On hiking tours in the Swiss mountains, one often sees landslide debris at the foot of a mountain front, and landslide scarps where the debris originated. If you draw a line from the toe of the debris to the top of the scarp, you may find, as Albert Heim(1882) and his colleague Eugene Müller-Bernet did, that the line has a slope angle of about 30°, just like the critical angle of repose of a stone.

When a mountain falls, it does not tumble down like a rolling boulder, it breaks into thousands or millions of pieces and flows. The motion of the famous 1881 Elm landslide as a three-act theatrical spectacle (Fig. 5.1) was described by eye witnesses to Reverend Buss in 1881:

"When the rock began to fall, the forest on the falling block moved like a herd of galloping sheep; the pines swirled in confusion. Then the whole mass suddenly sank...

Then the fallen mass hit the flat floor of a slate quarry and completely disintegrated. I (teacher Wyss) saw the rock mass jump away from the ledge. The lower part of the block was squeezed by the pressure of the rapidly falling upper

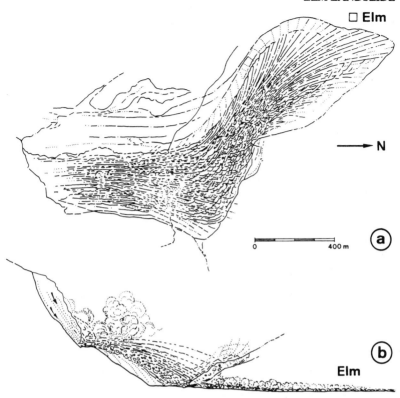

□ **Elm**

⟶ **N**

0 ———— 400 m

ⓐ

ⓑ

Elm

Fig. 5.1. Elm rockfall, 1882 (after Heim 1882). **a** A map showing the lines of movement of the Elm rockfall debris. The main mass turned to flow in a northwesterly direction down the valley toward the village. **b** A profile showing that the rockfall debris hit a rocky ledge on its steep descent (Act I) and cascaded like a waterfall (Act II), before the main mass speeded away like an express train in the direction of Elm (Act III). A small part surged on its own momentum 100 m up the hill, overtaking and burying numerous persons in their vain attempts to flee

part, disintegrated and burst forth into the air ... The debris mass shot with unbelievable speed northward toward the hamlet of Untertal and over and above the creek, for I could see the alder forest along the creek under the stream of shooting debris...

Once the driving stream of debris reached the bottom of the slope, a branch was sent directly north and surged up the side of the valley to a height of about 100 m. The main surge, however, was deflected toward the northwest and went down the Sernf Valley. The debris mass did not jump, did not skip, and it did not fly in the air, but was pushed rapidly along the bottom like a torrential flood. The flow was a little higher at the front than in the rear, having a round and bulging head, and the mass moved in a wave motion. All the debris within the stream rolled confusedly as if it was boiling, and the whole mass reminded me of boiling corn stew... the smoke and rumble was terrifying. I (11-year old Kaspar Zentner) now ran breathlessly over the bridge and bent around the corner of Rudolf Rhyner's house. Then I turned back and held myself firmly against the house. Just as I went past the corner, the whole mass shot right past me at a distance less than 1 m away. The debris flow must have been at least 4 m high. A single step had saved me. During the last jump, I noticed that small stones were whirling around my legs like leaves in autumn wind. The house crunched, moved and seemed to be breaking apart. I fled on hands and knees and no stones had hit me. I did not feel any particular air pressure. I went back to my home at Müsli down the street and found it in ruins, some 80 steps from its original position. Later I retraced my steps from the spot where I first saw the rock begin to fall to the corner of Rudolf Rhyner's house, and found the distance measured 200 to 300 steps. The time of my running is estimated to be about 40 seconds."

During these 40 seconds, the debris traveled about 2 km. The motion of the Elm debris has been described by phrases such as "round and bulgy head", "wave motion", etc., likened to conventional characterization of debris flows. The motion has also been compared to "torrential flood" or "boiling corn stew", except the interstitial fluid between the colliding blocks was not water, but dry air and dust. Debris flows and torrential floods are both fluid-gravity flows, but the debris flow of Elm was a sediment-gravity flow, and one that traveled at a catastrophic speed. It has been called *sturzstrom* or catastrophic debris flow to distinguish it from mudflows.

In contrast to the English usage of landslide, Heim stated that rock falls and rockfall debris flow. The Elm debris did not slide; they flowed. Sediments could flow if they are *fluidized* or *liquefied*, namely, when solid particles of a sediment-fluid mixture have been dispersed to such an extent that the individual grains are no longer in contact. The weight of fluidized solid is supported by a fluid force, namely, the inertial force of the rising fluid. Avalanche debris flow, but they are not exactly fluidized, because the weight of debris is not supported by fluid force. In fact, the flow of the avalanche debris is dependent upon grain contact in the form of collision, as described by Heim:

"Where a large mass, broken into thousands of pieces, falls at the same time along the same course, the debris has to flow as a single stream. The uppermost block, at the very rear of the stream, would attempt to get ahead. It hurries but strikes the block, which is in the way, slightly ahead. The kinetic energy, of which the first block has more than the second, is thus transmitted through impact. In this way the uppermost block cannot overtake the lower block and has to stay behind. This process is repeated a thousandfold, resulting eventually in the preservation of the original order in the debris stream. This does not mean that the energy of falling blocks from originally higher positions is lost: rather the energy is transmitted through impact. The whole body of the avalanche debris is full of kinetic energy, to which each single stone contributes his part. No stone is free to work in any other way."

Heim wrote this in 1932; he could not have known that the movement he postulated would be described some 20 years later by Bagnold(1962) as the turbulent flow of a dispersion of cohesionless grains. Bagnold used the term *grain flow* to describe the flow of concentrated grain dispersions in a flowing fluid medium. The grains are not transported by the fluid stress exerted by the interstitial fluid, the tangential force to move the grains consists largely of components of the effective weight of the grains themselves. Bagnold reasoned that a granular mass cannot flow without some degree of dispersion. In the case of grain flow, the dispersion must be upward against the downward body force. He demonstrated by experimentation the existence of such a dispersive stress, normal to the flow direction of the grains. He pointed out further that this stress is not a fluid stress, but originates from grain collisions in a dispersion of high concentration :

"When the grains are but a diameter apart or less (volume concentration 9% for spheres) the probability of mutual encounter, always finite for concentrations of a random grain array undergoing shear, approaches a certainty. The grains must knock or push each other out of the way according to whether or not the effects of their inertia outweigh those of fluid's viscosity, and both kinds of encounter must involve displacement of grains normal to the planes of shear. So it seems probable that the required normal dispersive stress between the sheared grains might arise from the influence of grain on grain. Further, the same encounters should give rise to an associated shear resistance additional to that offered by the intergranular fluid."

Comparing the passage quoted from Heim with that from Bagnold, it is clear that Heim's postulate for the physics of the motion of the avalanche debris is similar to that envisioned by Bagnold for grain flow. The only difference is the force applied to induce horizontal motion. For rockfall debris, the driving force is the weight of the debris. For the grain flow of Bagnold, the driving force is the fluid force of the overlying fluid, which exerts a shearing stress on the grain flow.

Streams flow downslope. The potential energy is wholly converted to kinetic energy, if there is no friction. In fact, friction exists, and the total energy of a mass is reduced by friction as it flows downstream. Using any level, such as sea level, as a reference, one can trace the sum of the actual potential energy and the kinetic energy equivalent of potential energy of a flowing mass as it moves downstream (Fig. 5.2). The plot of the decreasing energy with distance is called the *energy line*, which is a straight line with a negative slope if friction is the only resistance.

Rock falls, and the fallen mass drives down the *Fahrböschung* with such speed and momentum like a motor-driven vehicle. The line connecting the top and the toe of the driving slope is the so-called energy line of a sediment-gravity flow: Fig. 5.2 shows that the potential energy variation is shown by the line APP'E for the main mass and the line APP'E' for the surge. Theoretically, each stone with mass m reaching point P has lost its potential energy by $m g h_2$ or NP, of which a part (NO) is dissipated by friction and another part (OP) is converted into kinetic energy. At point P", the potential energy loss is converted to friction N"O" and kinetic energy O"P". The subsequent surge up to E' converts a part of the kinetic energy into the potential energy gain $m g (h_1 - h_2)$. The main mass flows with decreasing velocity until all the kinetic energy is consumed at point E. The energy line indicates the sum of kinetic and potential energy at any point of travel, and is a straight line with negative slope (AE), because the rate of energy dissipation is assumed to be constant.

Heim and Müller found that the slope of the energy line for smaller landslides is about the same as the coefficient of friction of sliding blocks. They referred, therefore, to this slope as the *equivalent coefficient of friction*, and the slope angle is about 30°.

Unlike the coefficient of friction, which is a material constant and whose value is independent of the size of a sliding block, the value of the apparent coefficient of friction is a function of the total volume of a fallen mass. Heim first made the observation that the slope of the energy line is smaller for larger rockfalls. I have compiled some of the observational data in Table 5.1, and plotted the data in Fig. 5.3.

The small friction of larger rockfalls led unexpectedly to an excessive distance of debris transport. This phenomenon is the cause of many natural catastrophes, whereby thousands of village inhabitants in a valley have been

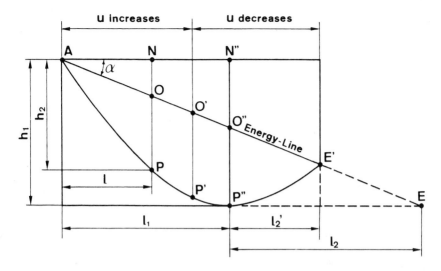

Fig. 5.2. Energy conversion during rockfall (after Heim 1932). The diagram shows that the conversion of potential energy into kinetic energy during a rockfall is accompanied by frictional loss (see text for explanation)

Table 5.1. Relation between friction and volume of rockfalls. The slope angle of energy line of small rockfalls is the same as the friction coefficient of rocky materials. This angle is, however, much smaller for rockfalls of very great volume

Rockfall	Fahrböschung angle	Apparent coefficient of friction	Volume of fallen mass
	(α, degrees)	($\tan \alpha$)	(10^6 m^3)
Airolo	33	0.65	0.50
Monbiel	23	0.42	0.75
Elm (60° turn)	16	0.29	10
Frank	14	0.25	30
Goldau	12	0.21	30-40
Kandertal (30° turn)	11	0.19	140
Flims (90° turn)	8	0.14	12,000

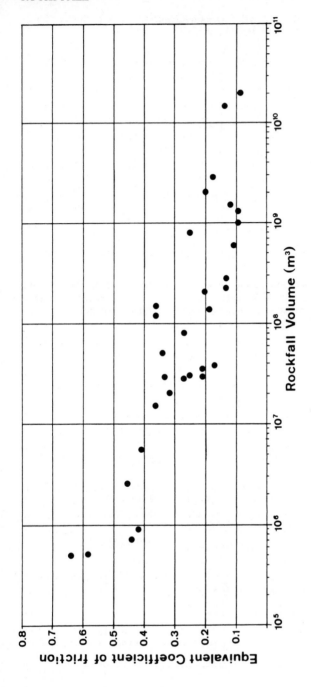

Fig. 5.3.

buried in rock avalanches. The longest avalanche debris stream traveled more than 700 km on the Moon and more than 2000 km on Mars.

We cannot use the Newtonian First Law of Motion to analyze the movement of avalanche debris, because the velocity of motion is not constant. On steep slopes, a moving mass accelerates when an excess of potential energy is converted into kinetic energy, far too rapid to be balanced by the frictional loss. After the catastrophic debris stream reaches the flat valley bottom, it moves on its own momentum. With little or no new supply from potential-energy conversion, but a steady attrition by friction, it would "lose steam and come to an abrupt end". Like a laminar flow, as the debris current stops, the bottommost debris reach a standstill first, while the overlying debris roll over at a lazy speed (Fig.5.4).

The sudden termination of the movement of the Elm landslide debris has been described (Heim 1882) by a 93-year old inhabitant of the village, who was inside one of the last houses struck by the debris but survived to tell the tale:

" I stood at the kitchen door, which was also the house door, and heard and saw with fear how the mountain came down. I thought that my wife was with our son next door, and wanted to go there to fetch her. But then the house crashed down, and I was caught by the wind and skidded back into the kitchen. I suddenly realized that I was rooted to the spot where I stood... I don't know how it came about and I was buried standing up to my neck among the broken lumber and stones. I could not move my arms and legs and was tortured by the extreme anxiety for my wife. After a long and horrible wait I finally heard the voice of my son. ' Is nobody here?' 'Oh, yes, Sepp', I shouted, ' I am here.' I was pleased that somebody else was also alive. Then my son started to dig me out. "

The furiously raging avalanche, which shot past the Rhyner house like an express train, came to a sudden halt within a distance of a few hundred meters. The sluggish movement of the last moving debris in laminar motion was not sufficiently violent to harm a frail old man, who was slowly buried in a stream of stones as if caught by a crawling mudflow.

Large rockfalls are commonplace, even those with a volume of more than 10 million m^3 have been happening every decade or so. A similar

---

Fig. 5.3. Relation between friction and volume of rockfalls. The inverse relationship between the apparent coefficient of friction and volume of rockfalls is shown by this diagram

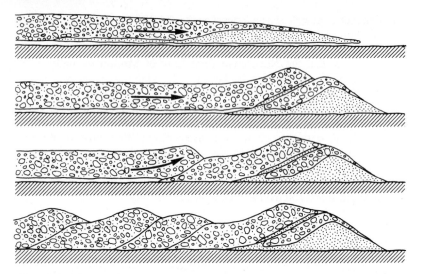

Fig. 5.4. Termination of movement of avalanche debris (after Shreve 1968). A distal rim is formed on initial impact. The successive piling up of late arrivals caused the formation of transverse ridges on the surface of, and imbricate structure within, avalanche debris

phenomenon is icefall, when a giant block of ice falls off the front of a glacier to generate streams of ice debris. The icefall of the Allalin Glacier some 30 years ago killed more than a 100 workers in Switzerland. An even greater catastrophe was the Huascaran event. In 1970, a mixture of glacial ice and rock basement fell down from a 6000-m peak of the Peruvian Andes. The rock-icefall generated an avalanche, which had a volume of about 10 million m^3. The catastrophic debris flow traveled 16 km and the speed reached a maximum of 280 km/h, faster than even the "bullet trains" of Japan. The avalanche caused an estimated 18000 casualties, mostly buried in the city of Yangay under the rockfall debris.

Rockfalls can also take place on submarine escarpments. In fracture zones associated with *transform faults,* the debris at the foot of an escarpment is commonly composed of angular blocks of fallen rocks. They are probably the deposits of small rockfalls. Deep-sea drilling in the Atlantic has extended into such breccia in the Romanche Fracture Zone.

This type of breccia is commonplace in the geologic record. The Jurassic formations of the Alps, for example, are noted for the occurrence of deep marine breccias. Numerous faulted troughs came into existence during the Mesozoic as a consequence of extensional tectonics. Breccias were deposited at the foot of fault escarpments. Some are *clast-supported;* the fragments in such breccias are all in contact so that the weight of each is supported by the other. These breccias have evidently been moved by clast collision in a sediment-gravity flow before their final deposition. Others are mud-supported; the fragments are embedded in a mud matrix. They were deposited by mudflows.

We tend to think that sediment-gravity flows should not travel very far. Rockfall debris are commonly deposited at the foot of mountains, and submarine breccias are rarely found far beyond the base of rocky escarpment. Yet avalanche debris of very large volume have traveled surprisingly great distances. The toe of the Saidmarreh landslip of Iran, for example, lies more than 20 km away from the top of the scarp, and the forward portion of the 20 km^3 debris climbed a 600-m high hill on its way before coming to rest in the next valley. Could submarine sediment-gravity flows also travel very large distances? What lesson have we learned from the Grand Banks event?

We have no eyewitness of the Grand Banks submarine slide, but the event was recorded by the breaks of the trans-Atlantic telegraph cables. Several oceanographic expeditions planned sampling by piston-coring of the debris deposited by the Grand Banks event. They were able to sample turbidites at distant sites on the abyssal plain, but had difficulty in penetrating the bottom deposit at the foot of the steep slope near Cable H (Fig. 5.5).

Seismic surveys during the 1980s, by David Piper(1985) and other scientists, revealed the presence of dune-like features on the floor of valleys descending from the shelf of the Grand Banks area down to great depths. These features first appear at 1500-m water depth and are still present at 4300 m, but they are not observed at 4600-m depth. The waves typically have wavelengths between 30 and 80 m, and an amplitude between 5 and 10 m. The slope of the "waves" facing downcurrent ranges from 25° and 45°.

Eventually a submersible, namely, a deep diving submarine, was sent down. To the astonishment of everyone, boulders up to 1 m in diameter were found in the valley. The gravels were derived from glacial outwash deposits which had been laid down, during the Ice Age, on the continental shelf offshore from Newfoundland. The dune-like features are thus called "gravel waves".

John Clarke reported in 1987 that the gravel deposit varies in thickness from 0.5 to 3 m, and consists of boulders, clast-supported pebble or cobble gravel, or mud-clast rich pebble gravel (Fig. 5.6). The upper surface of "gravel waves" is thinly draped by a normally graded, sandy or muddy sediment, obviously deposited from a suspension.

We all agree that there was an avalanche of shelf sediments from the Grand Banks down to the deep, and it was commonly assumed that the avalanche turned into a turbidity current which transported sands and gravels from the Grand Banks to the abyssal plain.

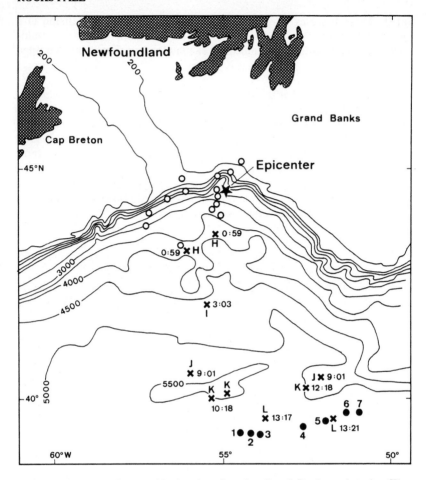

Fig. 5.5. Timing of the cable breaks after the Grand Bank earthquake. The cables are designated *H - L*. The *open circles* show the spots where the cables broke at virtually the same time as the earthquake, probably by a sliding movement. The *crosses* show the spots where the cables broke at the time given (in h) after the earthquake. The *solid circles* are stations where piston cores have been obtained by scientists of the Lamont-Doherty Geological Observatory; turbidites were sampled

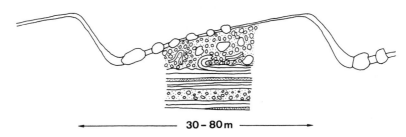

Fig. 5.6. "Gravel waves" produced by the Grand Banks event (after Clarke 1988). Chaotic gravels are found in channels leading from the Grand Banks area to the Sohms Abyssal Plain. The undulating upper surface is draped by fine detritus settled from a suspension. The so-called gravel waves may in fact be transverse ridges shown in Fig. 5.4

David Piper, Larry Mayer, and their colleagues(1985) from the Dalhousie University, Canada, considered the "gravel waves" a dune-like bedform. Such an interpretation, as we shall discuss later, implies that the coarse debris were transported as a bedload by a turbidity current, and that the "waves" moved like sand dunes. The absence of cross-bedding typical of dune deposits belies this interpretation. I suggested to Mayer, in an oral discussion, that the gravel deposit is not a turbidite but a sediment-gravity flow; the dune-like features are not a bedform, but are comparable to transverse ridges, marking the surface of an avalanche deposit. I came to this conclusion after an analysis of the physics of motion.

Kuenen assumed that all the cables were broken by a turbidity current. He used Chezy's equation to relate the speed to the postulated size of the Grand Banks turbidity current. The turbidity current had supposedly a density of 1.6 g/cm^3 and was flowing at more than 100 km/h when it broke the Cable H at the base of the continental slope (see Fig. 5.5). But a sediment-water mixture at such a high density cannot flow as a suspension; it would either move slowly as a submarine mudflow or very rapidly as a sediment-gravity flow. Accepting the estimated speed as evidenced by cable breaks, the latter alternative is the only viable postulate. Francis Shepard(1967), in fact, did point out that the cable-break data could be interpreted through the assumption of a turbidity current, originating at a point about 200 km from the Grand Banks. Such a current flowed at a constant speed of 7.7 m/s from somewhere downslope Cable H to an area beyond cable L (Fig. 5.7) He also envisioned that the Cable H was broken by a different kind of downslope movement, moving at much greater speed.

Using the Elm rockfall as a model, I can envision that a stream of gravel driving down the slope after a huge submarine slide was triggered by the

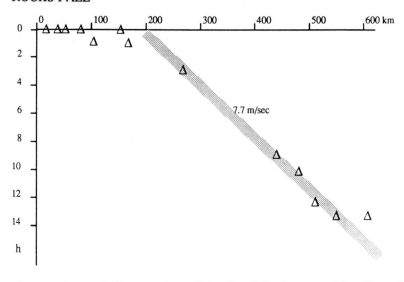

Fig. 5.7. Shepard's interpretation of the Grand Banks event (after Shepard 1967). Francis Shepard suggested that the cables *I, J, K, L* were broken by a turbidity current which flowed at a constant speed of 7.7 m/s. The Cable *H* was broken by a downslope movement of much greater speed, which I now interpret as a sediment-gravity flow

1929 earthquake. Each boulder, in its unsuccessful attempt at overtaking, collided with the one ahead and imparted its kinetic energy to speed up the motion of the "slow poke". The stream of gravel accelerated as it descended and may have attained speeds of hundreds of kilometers per hour. Even as the boulder stream slowed down, when it arrived at the bottom of the steep continental slope, its speed was still about 100 km/h. Suddenly all the gravels were deposited, while finer debris remained suspended in seawater to form a turbidity current which traveled several more hundred kilometers to be deposited on the abyssal plain.

This scenario suggests that the trans-Atlantic cable near the base of the slope was not broken by a turbidity current, but by a high-speed, sediment-gravity flow. If so, the coarse debris must have traveled at an average speed of 200 km/h before the Cable H was broken, about 1 h after the earthquake (see Fig. 4.2). Only the more distant cables were broken by a turbidity current.

Is it possible to calculate the speed of a sediment-gravity flow? Can we use Chezy's equation?

We have shown that Chezy's equation is derived from the consideration of the equilibrium of forces of an open-channel fluid flow. A sediment-gravity flow, however, hardly ever moves at a constant speed. The flow accelerates on its steep descent, but quickly slows down after it reaches the flat bottom. Chezy's equation is, therefore, not applicable for an analysis of the sediment-gravity flow of the Grand Banks event. We have to approach the problem through a comparison with the Elm rockfall.

The sediment-gravity flow at Elm has been observed and its movement timed by a teacher who made the observation from his apartment at the village of Elm. The estimated speed has further been checked by testimonies of other eye-witnesses. Heim and Müller-Bernet considered the physics of the motion, and suggested that the dynamics of the avalanche could be simulated to the sliding movement of a block down an inclined plane. From a consideration of equilibrium of forces

Gravitational force =   inertial force + friction, or:

$$m g \, \sin \theta = m a \, + \mu m g \cos \theta . \qquad (5.4)$$

The acceleration of the Elm rockfall should thus be

$$a = g \; (\sin \theta - \mu \cos \theta) . \qquad (5.5)$$

The value of μ, the equivalent coefficient of friction of the flow, is known, and the angle of the slope is also known, the acceleration at any point can thus be calculated. The acceleration is positive where the slope is greater than the gradient of the energy line, and negative where the slope is less. Substitute the calculated values of acceleration; the velocity at any point can be calculated on the basis of

$$u = u_o + a t . \qquad (5.6)$$

The results of the calculations by Heim and Müller are shown in Fig. 5.8.

It should be recognized, however, that sediment-gravity flow is not frictional sliding: When a boulder rolls down slope, the center of gravity of the boulder moved from the top of the slope to the point where the movement of that boulder stops. For rockfalls, however, the center of gravity of the fallen mass is not exactly at the top of the scarp, it should lie somewhere below the top. Also, the center of gravity of the debris does not lie at the toe of this catastrophic rockfall debris flow. None of the bolders of a rockfall run from the top of the scarp to the tip of the rockfall debris deposit.

The slope of energy line is thus not a coefficient of friction. It has been called *Fahrböschung* (runout slope) by Heim. As I have explained previously, this dimensionless number, like the coefficient of friction, is a measure of energy attrition during the movement (see Fig. 5.2). The

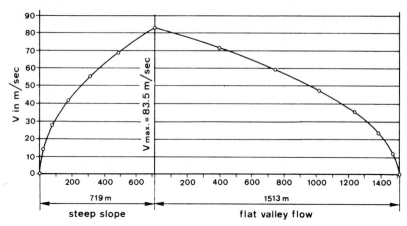

Fig. 5.8. The computed velocity profile of the Elm sturzstrom (after Heim 1932)

Fahrböschung, or the apparent coefficient of friction, of rockfalls with a volume smaller than 10^7 m^3 has values nearly identical to that of rock sliding, which is 0.6. The slope angle of the energy line of smaller rockfalls is thus about 30°, or the angle of repose of loose sediments in air. The values for very large slides have, however a considerably smaller angle for the energy line. This angle is only 17° for the Elm landslide, giving an apparent friction coefficient of 0.3, or about half that of the coefficient of sliding friction. The angle of the energy line of the great Saidmarreh landslip, which has a volume of 2×10^{10} m^3, is as low as 4°, and its apparent coefficient of friction is only 0.08 (see Table 5.1).

The angle of repose of water-saturated sediments is less than that of the dry sediments. Hydrostatic pore pressure reduces effective normal pressure and consequently friction. The subaqueous slide of Lake Zurich near Horgen in 1875 took place on a slope of less than 20°. The slope angle necessary to initiate a sliding movement is even less if the subaqueous sediment is clay-rich and thixotropic. The subaqueous slide on the shore of Lake Zug in 1887 took place on a subaqueous slope of less than 3°. This consideration explains why the Grand Banks submarine slide was initiated on a submarine slope of a few degrees only.

Once the motion was started, the potential energy of the moving mass was converted into kinetic energy to generate a sediment-gravity flow of boulders, cobbles, and gravels. The energy line of the Grand Banks sediment-gravity flow is the line connecting the top of the breakaway scar at Grand Banks to the tip of the boulder deposit at the foot of the continental slope.

For the sake of calculation, we can assume the termination of the sediment-gravity flow at some distance after the break in slope, at Cable H, which broke 59 minutes after the earthquake (see Fig. 5.5) The slope of energy line is

$\tan \theta$ = the depth difference divided by the distance between Grand Banks and Cable H
= 2400 fathoms/ 110 nautical miles
= 0.022 .

The steep continental slope south of Grand Banks has an inclination of about 2° or an average gradient of 0.035. The sine of the slope angle is thus also 0.035, and the cosine is almost unity. Substitute these values to Eq. (5.5) to calculate the acceleration, thus

$$a = 981 \text{ cm/s}^2 \text{ x } (0.035 - 0.022) = 13 \text{ cm/s}^2 .$$

Substitute this value into Eq. (5.6) to calculate u, we thus obtain a maximum speed of 40 m/s some 50 min after the earthquake at the slope break. From this point the slope is about 1:100, we should thus have a negative acceleration of 12 cm/s^2 for the next 10 min. The speed of the sediment-gravity flow at the Cable-H break should thus be 7 m/s less than that of the maximum speed, or 33 m/s. The calculated speed can be compared to the observed speed of 28 m/s (see Fig. 4.2). The agreement between the theoretical and observed values is amazingly good, especially in view of the fact that we would get irrational results, if we make the calculations by assuming the termination of the sediment-gravity flow at a more distant point.

For example, if we assume the sediment-gravity flow was terminated at Cable I, which broke 3:03 h after the earthquake, we would obtain a value of 0.011 for tan θ, and an accleration of 24 cm/s^2 to reach a maximum speed of 150 m/s at the slope break. The deceleration after this should be 10 m/s. The current should thus be reduced to 90 m/s at Cable H, but to a negative speed at Cable I. The irrational results prove that the assumption of a more distant termination is wrong.

This analysis suggests that the Cable H was broken by a sediment-gravity flow, which still had a speed of 28 m/s, or some 110 km/h when it approached the base of the slope. The avalanche debris came to a sudden halt somewhere between Cables H and I. Only the more distant cables were broken by a slowly decelerating turbidity current.

Sudden termination of avalanche movement causes the formation of a distal rim. Successive arrivals of waves of debris, piling up behind the rim, lead to the formation of transverse ridges and imbricate structures (Fig. 5.4). I concluded, therefore, that the "gravel waves" found by Piper and others are not a bedform, but piled-up structures like the transverse ridges. Thus in the future use of submersibles it is advised to search for imbricate structures within the gravel deposit.

The gravel avalanche of the Grand Banks is not a unique event. Geophysical investigations of a submarine canyon in the Gulf of Lyon in the south of France have revealed another instance of a sediment-gravity flow, which was generated by a submarine rockfall. Tertiary chalk blocks from the breakaway rim have been sampled at more than 2 km depth at the mouth of the canyon. The same type of "debris waves" as that found in the area south of the Grand Banks has been discovered in the Gulf of Lyon region.

Are all the breccias and conglomerates in the geologic record deposited by flows of such tremendous speed? Certainly not. They could be deposited by slowly moving fluid-gravity flows of high density, called debris flows. Mudflows are the most familiar type.

Sediment-gravity flows can, in some instances, be converted into fluid-gravity flows. After the sediment-gravity flow of the Huascaran slide reached Rio Santa, for example, part of the debris extended up the other bank of the river. Much of the debris, however, was deflected down the river and gradually changed into a debris flow. The debris traveled the 15 km in 12 min at an average velocity of 60 km/h, and velocities of 30 km/h or less were recorded downstream. In fact, the subaerial mudflow discharged from the mouth of Rio Santa may have continued on down the steep slope of the Pacific as a submarine mudflow.

The debris-flow deposit of Lake Lucern in Switzerland is believed to have been deposited by the subaqueous continuation of a debris flow originating from Mount Rigi. One spring day in 1795, the loose debris at the foot of Rigi broke loose with a thunderous roar. The talus and moraine materials were piled up in a debris flow that was 4 m high along a 1 km wide front. Slowly, but relentlessly, the flow moved down the valley, the village of Weggis on the shore of the lake was overwhelmed. The movement was slow; it continued for about 2 weeks before the water-soaked debris were emptied into the lake. The 1795 deposit was identified from the cores we obtained from Lake Lucern. The maximum thickness is 10 m at the base of the slope. The sediments are mud, and they show structures typical of soft-sediment deformation, such as slump folds, fractures, etc. The thick mud deposit was topped by a thin, silty graded bed.

It seems that the debris flow from Rigi via Weggis was not dispersed during its journey to the bottom of the 150 m deep lake. The subaqueous mudflow, like its subaerial counterpart, was a fluid-gravity flow, because the moving force was not induced by grain collision, but by the weight of mud, which behaved as a viscous fluid.

Landslides or submarine slides are not necessarily rockfalls. The falling or sliding block may consist of unconsolidated sediments, gravels, sands, or muds. The Swiss used the expression *stone avalanche* to designate slide deposits originating from loose stones or gravels. The sediment-gravity flow generated by the 1929 submarine slide at Grand Banks, for example, is strictly speaking a submarine stone avalanche, originating from the flow of a disintegrated Quaternary gravel deposit on Grand Banks. Similarly, we could use the expression *sand avalanches* to designate slide deposits of loose sands and *mud avalanches* to designate slide deposits of unconsolidated muds.

Stone avalanches have been described in sedimentological literature, although they have not been called such. In 1955, when I investigated the Pliocene deep-sea sediments of the Ventura Basin, California, I recognized the presence of a special kind of conglomerate. They consisted of well rounded cobbles and gravels, with the clasts touching one another (clast-supported). They are distinct from the conglomerates of alluvial origin by its matrix: The conglomerates in our Molasse deposits, for example, have a matrix of coarse sand and fine gravel, but those from Ventura have a mud matrix. The Ventura conglomerates are similar to the gravel beds deposited by the Grand Banks event; they are deposits of stone avalanches.

Are there sand avalanches in the geological record? Considering the frequent snow avalanches in the Alps, I would imagine that sand avalanches should be common occurrences in regions of a steep coastline fringed by sand beaches. Yet, no sand avalanche deposits have been recorded in sedimentological literature. Thick sandstone beds are, however, not uncommon in deep-sea sedimentary sequences, but they have been called *fluxoturbidite* .

I always had difficulty understanding this term. Why fluxo? Why turbidite? What is the difference between a fluxo- and an ordinary turbidite. It seemes that this is another case when a geologist wanted to hide his ignorance behind an exotic name.

The term first appeared in an article published by Stan Dzulynski, a noted Polish sedimentologist, and his colleagues. When I visited Cracow in 1976, I asked Dzulynski to show me a fluxoturbidite. He took me to the foothills of the Polish Carpathians to face the prototype of fluxoturbidite. It was a sandstone bed some 5 m thick, intercalated in a deep marine sedimentary sequence. Other sandstone beds of the sequence show graded bedding and other structures typical of deposition by turbidity currents; they are turbidites. The "fluxoturbidite" is relatively well sorted and show none of the turbidite features. There are faint laminations here and there, suggestive of resedimentation of a previously deposited sand.

I asked Dzulynski what was the mechanism of deposition of a "fluxoturbidite". He answered:

"The sand came down a steep submarine slope like snow avalanches in your mountains."

"Then why do you not call them sand avalanches?"

"Yes, we did, but then Kuenen came. He was an international expert, and we were just provincials. We wrote an article of our Polish work with joint authorship. Kuenen insisted that the term "sand avalanche deposit" is too long a name, with too many genetic connotations. Fluxoturbidite is shorter, and can always be considered a descriptive name if our interpretation is wrong."

This is unfortunate, of course, because the expression "sand avalanche" tells everyone clearly the interpretation of the authors on the basis of their evidence; fluxoturbidite is just another new fancy word beloved by fuzzy thinkers in their pretence to be technical.

Since then, I have seen many sand avalanche deposits in various parts of the world. Their internal structures are similar to snow avalanches. Where the disintegration of the original sediments was complete, the avalanche deposit is a massive sand. The movement of sand debris must have been so rapid that fluids are trapped. When the very porous deposit became more compact, interstitial fluid, under pressue, escaped upward on its way back to the seafloor. I have seen in Eocene deposits of Japan fine examples of networks of fluid-escape tubes.

Sliding deposits consisting mainly of mud sediments are common in the sedimentological record. An actualistic example of subaqueous sliding has been investigated by my student Kerry Kelts (1978); he wrote:

The citizens of Horgen, a village on the west shore of Lake Zurich, were proud of their new raildroad station, ... a building rose where it was swampy lake shore, thanks to loads and loads of sand and gravel-fill. The station was festively dedicated on September 20, 1875, then the autumn rains came. Two days later, the lake shore around the station started to collapse. In another two days, the station house, together with 6560 m^2 of newly reclaimed land slid away, and a minimum of 100000 m3 of debris went down a 20-degree slope to the bottom of the 140-m-deep lake.

Unlike the Grand Banks event, the Horgen slide was not triggered by an earthquake, but induced by the weight of landfill material. The main mass of the slide consists of lacustrine mud. The slide debris are now found deposited on a subaqueous fan at the foot of the steep slope. The resedimented deposit shows a sequence indicative of the progressive disintegration of the slide material (Fig. 5.9). The base, division **a**, shows slump folding, with little or no dispersed material. This is overlain by a mixture of large mud fragments with slump folding and soft clasts ("mud pebbles"), division **b**. This grades upward to the mud-clast facies, division **c**. Where the

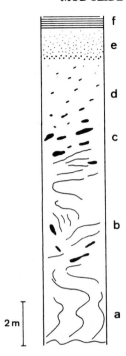

Fig. 5.9. Facies model of a subaqueous slide (after Hsü and Kelts 1985). Sediments redeposited after a subaqueous slide record the progressive disintegration of the slide mass, as shown by this idealized facies model. See text for definitions of divisions *a, b, c, d, e, f*

disintegration was partial, the original sedimentary structures, such as horizontal lamination or cross-lamination, are partially preserved. Since the sediment particles in both clasts and matrix are sand- or silt-sized, the boundary between the clast and matrix is commonly indistinct.

The overlying unit, division **d**, is a homogenized mud, which includes some mud clasts, but the boundary between such clasts and the homogeneous mud is barely visible. The highest sediment, division **e**, is not deposited by the sediment-gravity flow, but by a turbidity current; it consists of a fine-grained sand showing graded bedding. Division **f** is a normal lacustrine mud.

Mud avalanche deposits have been called slump or mudslide deposits in sedimentological literature. I see no need to introduce a new term. Such deposits are common in ancient deep-sea sequences. In a deep-sea drilling expedition to the South Atlantic, we encountered a 17-m-thick sequence at the foot of a fault scarp, deposited in a basin on the flank of the Mid-Atlantic Ridge. Slump deposits of lime mud or terrigenous mud are quite common in the Alps. One large deposit in the southern Alps is almost 100 m thick. In numerous instances, the slump deposit is overlain by a graded bed, which was presumably deposited by a turbidity current that had been generated by the slide.

Thick sedimentary breccias or slide deposits have also been called *olistostrome*. Single large blocks in sedimentary sequences are called *olistoliths*. The olistostrome concept was introduced by a field geologist, G. Flores, in 1955, when the fixistic school of tectonics dominated the geological thinking. Large displacement of rock bodies was believed to have resulted from downslope sliding under gravity. Seeing the tectonic melange of Sicily, which we now believe to be a mixture of brittle blocks and ductile matrix in a shear zone, Flores thought that it was an olistostrome, or a stratum of coarse debris, hence, the use of the word *strome*, deposited by viscous flow of mud under gravity. He used the prefix *olio-* (oil in Italian) to describe this sedimentary deposit, postulating slow motion like the creep of oil downslope. Olistoliths, in the opinion of Flores, are simply big slabs of rocks which somehow creep individually downslope to settle somewhere on deep-sea floor.

The so-called olistostromes have never been observed in an actualistic setting, although many parts of the ocean floor have been sampled by drilling during the deep-sea cruises of the last two decades. Deposits of slumps of sediment-gravity flows or mudflows are not common, but they are recognized as such. Thus there has been no need to use the term olistostrome to designate such deposits.

The practice of using the term olistostrome to designate mixtures of coarse and fine rock fragments is one of the worst examples of using the natural-history approach in science. A rock body was described, and was given an interpretation without citing an actualistic analogue for comparison. The postulated process has not been defined, nor is there a proper analysis of the physics of olistostrome motion.

In using the term, Flores implied that the debris moved slowly like the viscous motion of oil. If so, olistostrome deposits are debris-flow deposits; there is no need to replace a well-defined concept by a meaningless word. In fact, as I have discussed in this chapter, breccia debris in some instances have not creeped forward like viscous oil; they may have cascaded downslope as catastrophic debris flows. To use olistostrome to designate that kind of breccia would be a misinterpretation. Furthermore, many of the so-called olistostromes are not even sedimentary deposits. The mixture of blocks or slabs in a sheared matrix is now called melange or tectonic melange, which is formed in shear zones, such as subduction zones.

The term "olistolith" was used to designate a physically impossible process of moving single rock slabs across nearly flat deep-sea floor. The presence of large exotic blocks in a fine matrix is not an uncommon phenomenon, but we need not violate physical principles to explain this natural phenomenon. Olistoliths in tectonic melange are exotic slabs tectonically fragmented and incorporated in a matrix of ductile shear. Olistoliths in sedimentary deposits could be explained as a lag deposit, having been left behind after a submarine current has removed the finer debris of a rockfall deposit by erosion. I do not know of any examples which have to be explained by the physically impossible postulate of a slab creeping slowly and many kilometers down a submarine slope of a few degrees or less.

Suggested Reading

In the best tradition of the uniformitarianism and the natural-history approach to geology, one should study sediment-gravity flow deposits by reading the observations of actual rockfalls by eyewitnesses. For those who read German, the book by E. Buss and A. Heim *Der Bergsturz von Elm* (Zürich: Wurster and Cie, 163 p, 1881) is a must; Buss was a minister of the church and interviewed dozens of survivors of the catastrophe, and Heim was Professor of Geology at ETH and provided a scientific analysis of the natural phenomenon. A short article with the same title was published by Heim a year later (Dtsch Geol Ges Z 34:.74-115, 1882). A third work by Heim *Bergsturz und Menschenleben* (Zürich: Fretz and Wasmuth, 218 p, 1932) was written in his last years; it is an excellent treatise on the subject on downslope movements. A brief English summary of Heim's work is to be found in my article *Albert Heim: Observations on landslides and relevance to modern interpretations* in the book *Rockslides and avalanches*, edited by Barry Voight (Amsterdam: Elsevier, 833 p, 1978). The Voight volume gives an excellent account of these natural phenonmena.

Ron Shreve has made an important contribution to the study of landslides, and his dissertation *The Blackhawk landslide* (Geol Soc Am Spec Pap 108:47, 1968) was a medal-winning classic. Shreve and I disagreed, however, on the question whether a rock-fall slides or flows, and much of the idea contained in this chapter was my answer to Shreve, published in my article *Catastrophic debris streams (sturzstroms) generated by rockfalls,* (Geol Soc Am Bull 86: 128-140, 1975).

Francis Shepard discussed the Grand Banks event in the second edition of his *Submarine geology* (New York: Harper, 1967). The "gravel waves" of the North Atlantic were first described in an article by David Piper and associates on *Sediment slides and turbidity currents on the Laurentian Fan: sidescar sonar investigations near the epicenter of the 1929 Grand Banks earthquake* (Geology 13:538-541, 1985). The "gravel waves" deposit was the theme of a dissertation study (1987) by John Clarke of the Dalhousie University. The "debris waves" in a submarine canyon were described by Alberto Malinverno and W. B. F. Ryan when they gave a talk on *Large avalanche scars on the continental margin, Nice, France,* at the 1985 Annual Meeting of the Geological Society, and the manuscript is due to be published in a special publication of the Geological Society of America on *Sedimentologic consequences of convulsive geologic events* edited by H.E. Clifton. The facies model of subaqueous slide was discussed in the Hsü and Kelts paper on Swiss lakes, cited previously in Chapter 4.

The concept of "olistostrome" was presented by G.Flores as a discussion of a paper by F. Beneo *Les résultats des études pour la recherche pétrolifère en Sicilie* (Proc 4th World Petrol Congr, Sec 1:121-122, 1955). My critique of the concept was included in a short article *Melange and melange tectonics of Taiwan* (Proc Geol Soc China 31:87-92, 1988).

6 Suspensions Flow

Suspension Current - Auto-suspension - Bagnold's Criterion - Chezy-Kuenen Equation - Keulegan's Law - Energy Line

While rockfall debris move, little stones, sand and dust particles are "boiled" out of the moving mass and mixed with air to form a dust cloud or a suspension and this mixture of dust and air constitutes a *suspension current*. Stones and sand cannot remain long in suspension; they soon drop out and are deposited. Debris cones are thus often found on top of landslide blocks after the dust is cleared (Fig. 6.1). Like turbidites, the debris cones show graded bedding, indicative of sorting by size while settling from suspension.

When the streaming debris from the Elm rockfall came to a screeching halt, the overlying dust cloud swept past beyond the village as a suspension current and reached the village Matt more than 10 km away. Since the cloud was made denser than air by suspended debris, the current flowed under its own gravity, hugging the ground level. Smaller stones and coarse sand grains could not remain long in suspension; they settled out of the suspension first. Finer sand, silt, and clay-sized particles settled later. Eventually all suspended particles dropped out, as the turbulence of the suspension died down. The successive deposition of increasingly finer and finer debris from the suspension downstream from Elm resulted from a lateral sorting of the suspended materials during the transport, and consequently a lateral sorting of the grain size of the deposit: Whereas small stones and coarse sand grains are present at the base of a graded bed on top of the landslide debris of the sediment-gravity flow, only a thin layer of fine dust was deposited at Matt down the valley.

After the avalanche debris from Huascaran of Peru reached Rio Santa, they mixed with the river water to form a debris flow, which eventually slowed down to less than 30 km/h. When this sediment tongue spilled into the Pacific and flowed down a submarine slope, the debris flow must have mixed with seawater to form a suspension. Turbidity currents could thus be generated by a subaqueous mudflow, as Hampton(1972) demonstrated with his flume experiments. I have no knowledge as to whether such a turbidity current deposit has been found offshore from the mouth of Rio Santa. We do know, however, that a turbidity current has been generated by the subaqueous continuation of the Rigi/Weggis mudflow in Lake Lucern, because a thin graded bed overlying the debris-flow deposit has been found by our piston-coring of the lake sediments.

Fig. 6.1. Debris cones (after Heim 1882). Debris cones are found on top of landslide blocks at Elm and on Sherman Glacier. The debris have settled out of suspension and the deposits thus commonly show graded bedding. The landslide blocks have been transported by sediment-gravity flow, the suspension is a fluid-gravity flow

Debris flow is a fluid-gravity flow, the fluid is mud which is so viscous that the sediment-water mixture tends to flow viscously. Turbidity current is also a fluid-gravity flow, but the mixing of debris flow and ambient water produces a suspension less dense and less viscous than mud, so that the current flows turbulently.

Turbidity currents are not only derived from debris flows, they may be derived directly from a submarine sediment-gravity flow. The sand, silt, and clay particles "boiled" out of a stream of avalanche debris from the Grand Banks mixed with seawater to form a suspension and this kind of suspension current in water, as postulated by Kuenen and others, flows downslope as a turbidity current.

After the termination of the Grand Banks sediment-gravity flow at the foot of the continental slope, sands and muds in suspension flowed downslope to break the cables which had been laid across the abyssal plain, as far as 800 km distant from the origin of the slide. At the most distal site of the cable break, the speed was reduced to less than 10 m/s. The deposition of a meter-thick turbidite at that site has been confirmed by coring. The sedimentological evidence thus confirms the postulate of the two stages of sediment transport: by sediment-gravity flow and by turbidity current. I have shown that Chezy's equation is not applicable to the analysis of the motion of sediment-gravity flow. Is it applicable to the evaluation of the speed of turbidity current?

A turbidity current is driven by the gravity of the suspension, which is related to the effective density of the suspension. Larger sediment particles settle rapidly from suspension. They tend to congregate at the bottom of the current and are dragged along by the fluid stress exerted on the grain flow by the overlying flowing suspension. The finer silt and clay particles remain in suspension, being uplifted by the upward component of the turbulence. The energy loss by a turbidity-current flow is, therefore, not only related to the resistance to fluid flow such as that considered by Chezy, additional energy has to be expended to maintain the falling particles in suspension.

In order to maintain sediment particles in suspension, not only the competence, but also the capacity of the fluid force must be considered. While the competence to transport the largest particle is, according to Shield's diagram, a function of the speed of fluid flow; the total capacity of sediment transport is related to the power of the current, i.e., available energy per unit time. Taking air transport as an example, we know that an airplane losing power has to dump part of its load to keep flying at the same speed. When a current loses its power, part of its suspension settles. The deposition of turbidite is, therefore, an indication that the current is no longer accelerating, but is losing power.

The power to keep a sediment load in suspension counteracts the power of the weight of the falling particles which is

$$\mathbf{P_S} = F_S \, s/t = \Delta\rho_S g \, V_S \cos\theta \, u_S \; , \tag{6.1}$$

where F_S is the effective weight of the sinking particles, $\Delta\rho_S$ their effective density, V_S their total volume and u_S is the average settling velocity of the particle, and s is the distance of settling per unit time.

The gravity of the falling particles can also be expressed by the effective weight of the suspension

$$F_s = \Delta\rho_s \, V_s \, g = \Delta\rho_c \, V_f \, g \; , \tag{6.2}$$

where $\Delta\rho_S$ is the effective density of the solid particles and $\Delta\rho_c$ the effective density of the current. Substitute Eq. (6.2) into (6.1), then:

$$\mathbf{P_s} = \Delta\rho_c \, V_f \, g \, u_s \; . \tag{6.3}$$

Another part of the current power is expended to overcome the power of the fluid resistance, which, as we have shown before, is

$$\mathbf{P_r} = F_r \frac{s}{t} = C_f \, \rho \, \frac{u^2}{2} \, A \, u \; , \tag{6.4}$$

where u is the velocity of the fluid flow, or the distance of flow per unit time.

The driving power is gravity, or

$$\mathbf{P_g} = \Delta\rho_c \, V_f \, g \sin \theta \cdot u \quad .$$

Equating the driving and resisting powers, we have

$$\Delta\rho_c \, V_f \, g \sin \theta \cdot u = \Delta\rho_c \, V_f \, g \, u_s + C_f \, \rho \, \frac{u^2}{2} \, A \, u \quad , \qquad (6.5)$$

where $d = V_f/A$. Divide Eq.(6.5) by $\Delta\rho_c \cdot V_f g \cdot u$, then

$$\sin \theta = \frac{C_f}{2\left[\dfrac{\Delta\rho_c \, g \, d}{\rho \, u^2}\right]} + \frac{u_s}{u_f} \quad . \qquad (6.6)$$

where d is the thickness of the current. The angle θ is the smallest angle of the bottom slope required so that the driving power of a turbidity current is sufficient to keep all sediment particles in suspension. This angle has also been called Bagnold's criterion for auto-suspension. The term $\rho \, u^2$ is an expression of inertial force, and $\Delta\rho_c \, g \, d$ a gravity force. The ratio

$$\left[\frac{\Delta\rho_c \, g \, d}{\rho \, u^2}\right]$$

or the ratio of the gravity force to inertial force has been defined as the *Richardson Number*, **Ri**. Substitute this expression into Eq. (6.6), then

$$\sin \theta = \left(\frac{1}{2}\frac{C_f}{R_i} + \frac{u_s}{u_f}\right) \quad . \qquad (6.7)$$

This equation states that the minimum angle of slope required to keep sediment particles with average settling velocity u_s in suspension is positively correlated to the resistance coefficient C_f and to the settling velocity. It is evident that the critical angle is small when only very fine silt and clay-sized particles are to be kept in suspension. The relation of the angle to fluid velocity is, however, not clear-cut, because u_f enters Eq. (6.7) both as a factor and as a quotient. We can appreciate that particles can be better suspended at great velocity because of greater turbulence, and we also understand that this greater turbulence also leads to more mixing of the current and the ambient fluid, causing the suspension to be diluted and thus to decay.

The value of the settling velocity can be estimated from Stokes' law and that of C_f is obtained from the Nikuradse or Moody diagram. Bagnold estimated that the angle should be less than 0.5° to maintain clay-sized particles in suspension. In other words, clay suspensions flowing down slopes greater than 0.5° would flow as a turbidity current (or auto-suspension) forever.

Bagnold's theoretical prediction was confirmed in 1978 by my student Stephan Lüthi in a laboratory investigation. Using a mixture of chalk powder and water, suspension currents were produced in flume experiments. The effective settling velocity of the suspended particles, which have an average diameter of about 3 microns, has been determined to be 0.004 cm/s. The suspension current has a speed of 12 cm/s and a Richardson number of 1.1. Substituting the experimental data into Eq.(6.6), Lüthi found that the angle for auto-suspension is 0.41 degree for his experimental mixture.

In the special case where the suspended particles are very small, the settling velocity term approaches zero, we have, therefore, from Eq.(6.6)

$$\sin \theta = \frac{\dfrac{C_f u^2}{2g}}{\left(\dfrac{\Delta\rho}{\rho}\right) \cdot d} \; . \tag{6.8}$$

Furthermore, for the very small angle θ, $\tan \theta$ is equal to $\sin \theta$. Equation (6.8) can thus be written as

$$u = \sqrt{\frac{2g}{C_f} \cdot \frac{\Delta\rho}{\rho} \cdot d \cdot \tan \theta} = \sqrt{\frac{\Delta\rho}{\rho} \cdot d \cdot s} \; . \tag{6.9}$$

This is the modified form of Chezy's formula used by Kuenen to relate the size of the Grand Banks current to the observed flow speed. In other words, we could indeed use Chezy's formula to evaluate the density-thickness factor of a turbidity current if the suspended particles are mainly of clay size. Serious errors are introduced, however, if sand- and silt-sized particles are kept in suspension.

Can the value of resistance coefficient as given by Moody's diagram be used? In the early 1970s, David Kersey and I did a series of experiments to evaluate the resistance coefficient to turbidity current flow. We found that the values of the coefficient for flows with **Re** values of 10^3 to 10^5 are not much different from that calculated on the basis of the experimentally determined friction factor (Fig. 6.2). Remarkable is the fact that the resistance is so small that turbidity currents flowing on slopes with an angle of 0.5° generate more power by gravity than the attrition of power by fluid resistance.

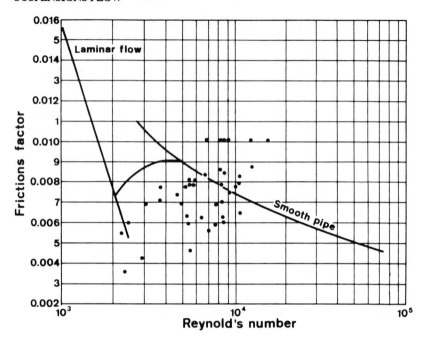

Fig. 6.2. Comparison of the friction factors (after Kersey and Hsü 1976). The *curves* show the variation of the friction factor **f** as a function of the Reynolds number as determined by water flow through pipes. The *points* show the experimental results of turbidity-current flow through a flume

The relative ease with which a current keeps its clay-sized particles in suspension and the very small fluid resistance explain the fact that mud turbidites are deposited far from the origin. In fact, even sand-sized particles have been transported by turbidity currents far from the source. The sandy graded bed deposited by the Grand Banks turbidity current on the North Atlantic abyssal plain was more than 800 km away from the source. The sand turbidites in Vema Trench of the South Atlantic as sampled by the Leg IV of the JOIDES Deep-Sea Drilling Project, have heavy minerals such as ruby and beryl which are derived from the Amazon region of Brazil more than 1000 km away.

When a current flows upslope, however, the driving power is switched off, and the attrition causes a rapid decrease in the current power. Suspensions may flow for long distances down a gentle incline, but tend to be ponded where the slope is reversed. Not knowing that, I was surprised by my discovery in 1955 that the coarse Pliocene turbidite sands of Ventura are not deposited at the foot of the continental slope nor on a submarine fan; they are present in the deepest depression of the Pliocene basin, where they constitute the *ponded facies*. This same pattern of turbidite distribution has been found by my student, Kerry Kelts. He studied the Lake Zurich turbidite which was deposited by a current generated by the 1875 subaqueous slumping near Horgen. The current apparently flowed for 7 km with little decrease in power, but much load was dumped where the bottom slope is reversed.

Suspension current is a special form of density current, whereby the density of the current is larger than the ambient fluid due to the suspension. Saline water also forms a density current, and the upstream invasion of seawater as a saline head is a very serious problem of coastal engineering. G. H. Keulegan of the U. S. Waterways Experiment Station carried out a series of studies in the 1940s and 1950s to study the movement of the saline head. By opening a lock, a saline solution (with effective density $\Delta\rho$) in a reservoir was released into a body of standing fresh water (with density ρ) in a horizontal flume. A density current of thickness d was generated. The relation obtained by Keulegan on the basis of his observations is

$$u = 0.71\sqrt{\frac{\Delta\rho}{\rho}g\,d} \quad . \tag{6.10}$$

Density currents in Keulegan's experiments flow in a horizontal flume or even upslope. The kinetic energy of the density current is converted from potential energy, but the center of gravity is not lowered by a downslope displacement of a current, but through the collapse of a column of dense fluid.

Consider the case of the collapse of a denser fluid of height H to form a density current with a thickness d. The potential energy before the "collapse" is

$$(\text{P.E.})_0 = 1/2 \;\; \Delta\rho_c \;\; V \; g \; H \;. \tag{6.11}$$

The potential energy after the "collapse" is

$$(\text{P.E.})_1 = 1/2 \;\; \Delta\rho_c \; V \; g \; d \quad . \tag{6.12}$$

The kinetic energy in the current after the collapse is

$$(\text{K.E.})_1 = 1/2 \; \rho \; V \; u^2 \quad . \tag{6.13}$$

Equating the potential energy loss to the kinetic energy gain, we have

$$u^2 = \frac{\Delta\rho}{\rho} g \; (\text{H} - d) \quad . \tag{6.14}$$

97

According to our observation, D is almost always half of H, we have thus

$$u = \sqrt{\frac{1}{2}\frac{\Delta\rho}{\rho}g\,d} = 0.71\sqrt{\frac{\Delta\rho}{\rho}g\,d} \quad.$$

This is Eq.(6.10) or the empirical relation found by Keulegan.

In the experiments carried out by Kersey and Hsü in a flume with a horizontal floor, the current thickness is slightly less than half of the height H. We have found that not all the potential energy lost has been converted to kinetic energy, a small part is lost as friction. We borrowed, therefore, the concept of energy gradient to analyze the mechanics of turbidity-current flow. This concept has been useful in the analysis of fluid flows, including not only those in open channels, but also sediment-gravity and groundwater flows. The gradient is defined as the change of energy per unit volume, also known as *specific energy*, with distance. Often it is convenient to express the total specific energy in terms of a potential energy, namely, the height of a head. The energy gradient can thus be considered an expression of the loss of this head with distance, as we have shown in the case of sediment-gravity flow (see Figs. 6.3, 6.4). This loss is induced by friction.

Using the concept of energy gradient, we have obtained numerical data to show that the friction loss of turbidity-current flow is indeed very small. A density current may thus flow down a very gentle slope of much less than 1° for a long distance until the density contrast between the current and ambient fluid is eliminated by mixing. These results can be considered as verification of Bagnold's analysis of the mechanics of auto-suspension.

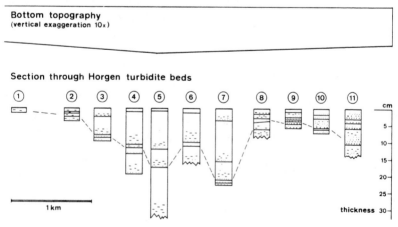

Fig. 6.3. Distribution of Horgen turbidity-current deposits (after Kelts 1978)

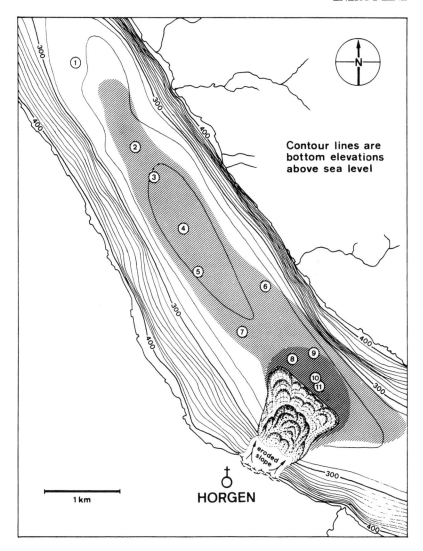

Fig. 6.4. Ponded turbidity-current deposits (after Kelts 1978). The thickness of a turbidity current deposit is related to the loss of transport power. This figure shows that the turbidity current deposits generated by the 1875 Horgen slides are thickest at Station 4, where the slope is reversed, resulting in a maximum loss of transport of power. *Contour lines* are bottom elevations in m above sea level and *circled numbers* are sampling sites

Suggested Reading

Since the idea was first proposed in 1950 by Ph. Kuenen and C.I.Migliorini in their paper *Turbidity currents as a cause of graded bedding* (J Geol 58: 91-127), the literature on turbidity currents and their deposits is voluminous. The development is a triumph of using the natural-history approach in sedimentology. The history of discoveries has been summarized by Roger Walker(1973) in his excellent review: *Mopping up the turbidite mess* (R. Ginsburg *Evolving concepts in sedimentology*, cited in Chap. 2).

I have chosen, however, to use the physical science approach to analyze the turbidity-current process in this chapter. One could start his further reading with Kuenen's 1952 article cited in Chapter 4. H. W. Menard reviewed in 1964 the various speculative hypotheses on the mechanics of the turbidity current (*Mar Geol Pacific*, New York: McGraw Hill). Gerald Middleton was the first within the earth science community to call attention in 1964 to a series of experiments made by G. H. Keulegan of the U.S. Army of Engineers. Keulegan had made the pioneering experimental studies on density currents with saline solutions and his conclusions are published in his progress reports to the National Bureau of Standards (e.g., First and Second Progress Reports on Project 48, 1946; 12th and 13th Progress Reports on model laws for density currents, 1957/58). Middleton experimented with suspensions and confirmed the applicability of Keulegan's law to turbidity currents. My associate David Kersey and I analyzed the theoretical basis of Keulegan's results, and investigated experimentally the energy relations of density-current flows (Sedimentology 23:761-789, 1976). H. A. Hampton (1972) made experiments to investigate *The rôle of subaqueous debris flows in generating turbidity currents* (J Sediment Petrol 42:775-793).

R.A.Bagnold defined the criterion for auto-suspension in his 1962 paper published in the Proceedings of the Royal Society (Ser A 265:315-319). Lüthi's experimental work was elucidated in his dissertation *Zur Mechanik der Turbiditätsströme* (Diss ETH 6258, 1978). Studies of turbidity-current deposition in Recent environment were the theme of a dissertation by Kerry Kelts *Geological and sedimentary evolution of Lakes Zurich and Zug, Switzerland* (Diss ETH 6146, 1978).

7 Sand Waves Migrate

Froude Number - Richardson Number - Model Theory - Bedform
Point-Bar Sequence - Facies Models

We have so far discussed two types of fluid forces, those varying linearly with velocity and those varying with velocity squared.

The first kind, as we discussed in Chapter 2, is exemplified by the viscous resistance between a slowly moving object and a fluid medium or that between a fluid and a pipe of small diameter. We are familiar with viscous forces because many liquids, such as heavy oil or dense mud, are sticky. The stickiness is an expression of the viscous resistance to movement, which is large for fluids of high viscosity. The second kind, or the inertial force, overcomes the inertia of a fluid, and gives kinetic energy to its movement.

When a liquid moves slowly, the viscous force is dominant, and the movement is laminar, but the fluid becomes turbulent when the inertial force is dominant. We have discussed Reynolds' discovery that speed is not the only factor, the diameter of the conduit and the viscosity of the liquid can also determine whether a flow is laminar or turbulent.These are the parameters which determine the mode of fluid movement. The combination of these parameters is a quality expressed numerically by the Reynolds number, which is a relative measure of the inertial and viscous forces.

In the last chapter, we introduced another number: the Richardson number. We cited Bagnold's work which stated that the angle necessary for auto-suspension is inversely proportional to the numerical value of the Richardson number: the critical angle has to be larger for current with a smaller Richardson number. What is the physical meaning of this statement? What is the Richardson number?

In fluid motion on Earth, gravity is ubiquitous. The gravity is a moving force of a river current, as expressed by the relation

$$F_g = m \, g \, \sin \theta \, , \tag{7.1}$$

where θ is the slope of the river channel.

When a river empties into a sea, this force will propel the river water to move some distance forward, until the kinetic energy of the river water is dissipated, or until the river water has lost its identity through mixing with sea water. Take, for example, the case when the Mississippi River flows into the Gulf of Mexico. The riverwater with a density of 1.00 g/cm^3 is lighter than the seawater which has a density of 1.03 g/cm^3. The riverwater spreads out to form a *plume,* or a fan-shaped layer of freshwater above the seawater. Of course, the plume does not extend to infinity, even though the Mississippi continues to empty into the Gulf. The plume is limited in extent, because the freshwater in the plume mixes with the saltwater. The mixing is done by bringing a volume of the lighter water in the plume downward (from the depth d_o to d) where it mixes with the seawater at the lower boundary of the plume. Work against gravity requires energy. The gravitational force opposing the mixing movement is

$$F_g = \Delta\rho \ V \ g \ d \ , \qquad\qquad (7.2)$$

where $\Delta\rho$ is the difference in density between the two fluids.

What provides the energy, or the force to induce mixing? The kinetic energy of the moving water in the plume is the only source, and this energy is expressed by the inertial force

$$F_i = \frac{1}{2} \ m \ u^2 = \frac{1}{2} \rho \ V \ u^2 \ . \qquad\qquad (7.3)$$

The tendency to mix is governed by the relative importance of the gravity and the inertial forces: the less the gravity force, the more the inertial force (derived from the kinetic energy), the more the mixing.

Tendency to mix = gravity force/inertial force.

Introducing a numerical expression as an abbreviation, now called the Richardson number

$$R_i = \text{dimensionless number} \ \frac{\text{gravity force}}{\text{inertial force}} = \frac{\Delta\rho V g d}{\rho V u^2} = \frac{\dfrac{\Delta\rho}{\rho} g d}{u^2} \ ,$$

we can state that the tendency to mix depends upon the value of the Richardson number; the larger the number, the less is the tendency of mixing.

Now going back to the criterion of auto-suspension, Eq.(6.7) states that the critical angle for auto-suspension is inversely proportional to the value of the Richardson number. The critical angle is small for a density flow with a large Richardson number. Such a two-phase flow with minimum

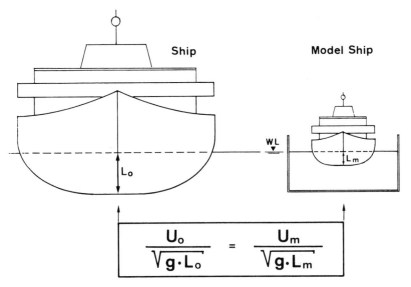

Fig. 7.1. Froude number. The Froude number is devised to test the dynamic similarity of a model to its prototype. u is the speed, L is a linear dimension; and g is gravitational acceleration. The *subscripts* o and m designate the parameters for original and model, respectively

turbulent energy has the least tendency for mixing at the boundary and could thus be considered more stable.

Aside from the Reynolds and Richardson numbers, the Froude number is a third dimensionless number that has been invented to characterize the dynamics of a fluid flow. What is the Froude number and who is Froude?

In the last century, when the British Navy was expanding, a prize was advertised to be awarded to the person who could devise a model boat which could predict the behavior of a real ship in motion. Froude was a person who built model boats, and he found that such a model could be made, but the model has to be geometrically similar, and its draft and speed are related to those of the vessel by the following relations (Fig. 7.1)

$$(\text{Speed}_{model})^2 / \text{draft}_{model} = (\text{Speed}_{ship})^2 / \text{draft}_{ship} . \qquad (7.4)$$

Froude discovered this secret through experience, but this relation can be derived from a consideration of the theory of models.

The theory requires that every quantity in a model must be properly scaled in order to achieve geometric, kinematic, and dynamic similarities to the prototype.

Geometric similarity is easy to visualize. If the parts of the model have the same shape as the corresponding parts of a prototype, the two systems can be considered geometrically similar. Similarity can be defined mathematically through the introduction of a concept called scale factor.

Let us designate the three dimensions of a prototype and its model by symbols x_o, y_o, z_o and x_m, y_m, z_m, respectively. The scale factors for lengths in the x, y, z directions, are

$$x_m / x_o = K_x$$
$$y_m / y_o = K_y \qquad\qquad (7.5)$$
$$z_m / z_o = K_z$$

and in any direction is

$$L_m/L_o = K_L .$$

A model is geometrically similar to the prototype when

$$K_x = K_y = K_z = K_L .$$

The model is geometrically distorted, when

$$K_x \neq K_y ; \text{ or}$$

$$K_x = K_y \neq K_z ;$$

the ratio K_y/K_x or K_z/K_x is called the distortion factor. In making models of landscape, the vertical scale is often distorted. A vertical exaggeration of ten times has a K_z/K_x value of 10.

The scale factor for time is a more abstract concept. It is defined by

$$t_m/t_o = K_t , \qquad\qquad (7.6)$$

where t_m and t_o represent the time elapsed or required for an event to be simulated in the prototype and in the model, respectively.

Since velocity is defined by distance per unit time, the scale factor for velocity is

$$K_u = \frac{u_m}{u_o} = \frac{\dfrac{x_m}{t_m}}{\dfrac{x_o}{t_o}} ,$$

$$K_u = K_x / K_t \text{, similarly}$$
$$K_v = K_y / K_t \quad (7.7)$$
$$K_w = K_z / K_t \quad .$$

The condition of kinematic similarity is, of course:

$$K_u = K_v = K_w .$$

Through the introduction of the scale factor for acceleration, the condition for a geometrically similar model to be kinematically similar is:

$$K_a = K_L / K_t^{\,2} = K_u^{\,2} / K_L \quad . \qquad (7.8)$$

Dynamics relates motion to force. Geometrically and kinematically similar models are also dynamically similar if the condition is defined by

$$K_F = K_M K_L / K_t^2 \quad , \qquad (7.9)$$

where K_F and K_M are scale factors for force and mass, respectively, and

$$F_m / F_o = K_F ; \qquad (7.10)$$

$$M_m / M_o = K_M \quad . \qquad (7.11)$$

In complex dynamic processes more than one kind of force is involved. Take, for example, the case of an open-channel flow, the forces are the gravity (F_g), or the *body force*, and the resisting viscous and inertial forces (F_v, F_i). Each of the forces have a scale factor, namely, K_{Fg}, K_{Fv} and K_{Fi}. To construct a dynamically similar model, all three scale factors of forces should be equal, or

$$(F_g)_m / (F_g)_o = (F_v)_m / (F_v)_o = (F_i)_m / (F_i)_o = K_F . \qquad (7.12)$$

This may not be physically possible. In practice, we could neglect the similarity of a force which is of minor importance. In the case of fluid flow with a large Reynolds number, the viscous force may be negligible. The model is approximately dynamically similar if

$$(F_g)_m / (F_g)_o = (F_i)_m / (F_i)_o \neq (F_v)_m / (F_v)_o \quad . \qquad (7.13)$$

By rearranging the terms in Eq.(7.13) , we have

$$(F_i)_o / (F_g)_o = (F_i)_m / (F_g)_m \ . \tag{7.14}$$

In other words, the ratio of the inertial force to gravity force of a model must be similar to that ratio of the prototype, if the viscous force of the movement is negligible.

The inertial force of an open channel flow, Eq. (3.7) is

$$F_i = \frac{\rho u^2}{2} A \ .$$

The body force, or the weight, of the fluid is

$$F_g = \rho V g \ .$$

The ratio of the inertial force to the gravity force is, therefore,

$$\frac{F_i}{F_g} = Z \frac{u^2}{g D} \ , \tag{7.15}$$

where Z is a dimensionless number and D is a linear dimension. For dynamical similarity, we have from Eqs.(7.14) and (7.15)

$$\frac{u_m^2}{g D_m} = \frac{u_0^2}{g D_0} \ . \tag{7.16}$$

Defining the *Froude number* we then have

$$Fr = \sqrt{\frac{u^2}{g D}} = \frac{u}{\sqrt{g D}} \ . \tag{7.17}$$

Comparing Eqs. (7.16) and (7.17), we have:

$$Fr_0 = Fr_m \ . \tag{7.18}$$

In other words, to make a dynamically similar model of open-channel flow, the Froude number of the model must be similar to that of the prototype.

Now we could return and analyze what Froude had done. If we divide both sides of Eq. (7.4) by the gravitational acceleration and take the square root of both sides, we have Eq. (7.18), namely:

$$Fr_m = Fr_0 \cdot$$

So Froude has found by experience that the Froude number of the moving model boat must be the same as that of the moving vessel, if the model is to simulate the wave resistance of the boat movement.

A ship moves on water, and waves are produced at the air-water interface. The wave resistance is related to the weight of the boat and the speed of its motion. In boat construction, we are not concerned with viscous resistance, but more with the size and the speed of waves and the wave resistance. Therefore, the similarity of the Reynolds number can be neglected (see Eq.7.13), but the similarity in the Froude number is critical in reproducing the actual wave motion by model studies.

In the case of a stream flowing in an open channel, waves could also be produced at an interface – the water-sediment interface – if the channel is underlain by loose sedimentary particles. The size and shape of the waves produced, if at all, depend upon the Froude number of the current.

One sees horizontal laminations in many sandstones. Such a laminated structure indicates that layer after layer was laid down on a flat bottom. Horizontally laminations are thus formed in sands deposited by currents flowing over a flat bottom.

But the bottom is not always flat. In a flume demonstration, you can see the formation of ripples and sand waves under the influence of stream flow. You may also see that the sand waves are obliterated when the current velocity increases or the flow depth decreases. The configuration of the channel bottom, also known as the *bedform*, has interested sediment-ologists, because the bedform could be fossilized as internal structures of current-deposited sandstones.

Cross-bedding is a feature formed where sediment was laid down on an inclined surface (Fig. 7.2). In my student days, we read descriptions of cross-bedded sandstones, but we did not quite understand why sand should be deposited on an inclined surface which has a dip ranging from less than 10° to more than 30°. After I joined the Shell Development Company at Houston, Texas, in 1954, I was to collect sand samples for mineralogical studies. Two colleagues, Hugh Bernard and Robert Nanz, helped me on my sampling trips. One day, I was sick with flu and could not go out. Bernard and Nanz went without me. They surprised me in the evening with a visit. Excitedly they told me that they had found "giant sand waves" on the banks of the Brazos River near Houston. Furthermore, they told me, cross-bedding is the characteristic internal structure of the "giant sand waves".

Bernard spent the next decade studying the Brazos River. Meanwhile, he learned that geologists from the U. S. Geological Survey at Denver were carrying out experiments in large flumes. They reproduced the sand waves on the bottom of flume like those on Brazos. It was clear that cross-bedding

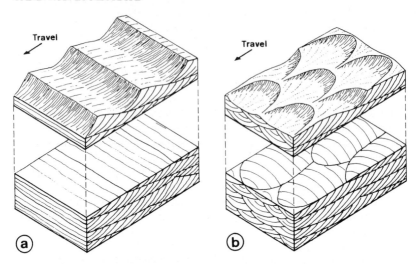

Fig. 7.2. Relationship of cross-bedding pattern to surface shape of sand waves (after Allen 1970). **a** Planar cross-bedding formed by migration of dunes or ripples with straight crests. **b** Trough cross-bedding formed by migration of sand or ripples with curved crests

was formed because of the migration of sand waves. Investigating the Holocene sediments by boreholes, Bernard and his colleagues found that the river valleys of the Gulf Coast are underlain by a sequence of alluvial sediments which he called the *point-bar sequence*. Eventually the point-bar facies model was proposed (Fig. 7.3).

The bottom of the sequence is a gravel deposit. The pebbles had been carried down during the Pleistocene when the streams on the coastal plains had steeper gradients. The sea level was lower during glacial stages because much of the water was locked in polar ice. The sediments have been reworked during the Holocene so that only pebble-sized clasts have remained. They are called a *lag deposit*; lag meaning that these coarse sediments have been lagging, or left behind by currents which sorted out and carried away sands and finer sediments; these currents were too weak to move the pebbles.

The gravels are overlain by cross-bedded sands. The sands are well sorted, because finer silts and clays have been carried away while the sands were being deposited. The power of the transport capacity of the Brazos floodwater at that time was enough to hold silts and clays in suspension, but not enough to overcome the settling power of sands (see Eq. 6.3). The cross-bedded sands are overlain by horizontally laminated sands. These have about the same grain size and sorting as the cross-bedded sands, suggesting

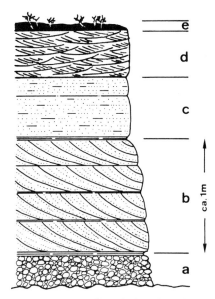

Fig. 7.3. Facies model of a point-bar sequence. **a** Gravel, lag deposit; **b** cross-bedded sand; **c** horizontally laminated sand; **d** cross-laminated silt: **e** mud. The sequence is about 3 m thick at Brazos River point bars, but over 10 m thick at Mississippi River point bars

similar transport power of the floodwater during the two stages of sand deposition.

Still higher in the sequence are cross-laminated silts. These poorly sorted sediments were deposited when the floodwaters were no longer powerful enough to hold all the silts and clays in suspension. The highest sediment of the sequence is a mud.

By repeatedly visiting a point bar on Brazos, near the city of Richmond, Texas, Bernard observed that the point-bar deposition took place during spring floods, which occurred once or twice a year after heavy rainfall. Throughout the year, the water of the Brazos River is clear, carrying hardly any sediment in suspension. The channel of the Brazos River near Richmond is underlain by lag gravels. The sands and silts on point bars originated upstream and were deposited during the floods. Cross-bedded sands were laid down first during an earlier stage when the point bar was submerged under deep floodwater. Horizontally laminated sands were deposited higher up on the point bar where the water depth was shallower. Still higher up, close to and on the natural levee of the river, are the cross-laminated silty sediments.

This sequence was first observed on the Brazos. Later, Bernard found a similar sequence of gravel, cross-bedded sand, horizontally laminated sand, and cross-laminated silt on the point bars of the Mississippi River, except everything is larger in scale. The gravels and sands are coarser, the cross-bedded and horizontally laminated beds are thicker. We now consider this association of sediments in this sequential order a signature of meandering-stream deposits.

Is there a theoretical basis for the point-bar model?

The model illustrates the sequence of deposition as the bedform of the point-bar surface under the floodwater, changing from sand waves to flat bottom and finally to ripples.

The changes of bedforms, as we mentioned, have been investigated experimentally and in the field. The bedforms are commonly related to the stream power (Fig. 7.4). Where the water depth of the flume is constant, higher stream power means higher current speed, or a larger Froude number. If this interpretation is correct, the change of bedforms of the Brazos point bar studied by Bernard is a manifestation of changing stream power.

With the help of my Chinese students He Qixiang and Zhao Xiafei, we have carried out flume studies here at ETH. Working in flumes with sandy and silty sediments, we found that the change of bedforms is not only related to current speed, and thus indirectly related to stream power, but is also governed by depth. Like the Reynolds number which governs the mode of fluid movement (laminar or turbulent), the Froude number determines the bedform at the sediment-water interface. Like Froude's model ship which sails with a dynamic similarity to the prototype, the sand waves are generated when the flume flow is dynamically similar to the stream current. This condition of dynamic similarity implies that the Froude number of the model and natural currents should be the same.

We found experimentally that sand waves begin to disappear when the Froude number of the flow exceeds 0.35 or 0.4. A wavy or rippled bottom changed to a flat bed as the water depth decreased, when the flow is characterized by a larger Froude number. We examined the publications by scientists of the U.S. Geological Survey, who recorded the change of bedform on the point bars of the Rio Grande, and found that the critical Froude number falls within the experimental range.

Since speed and depth are both factors governing bedform, the changes of the bedform on the Brazos can be interpreted as an interplay of the two. When floodwater first came, the Brazos channel was deep, and the velocity/depth ratio is such that the Froude number of the flow is sufficiently small to cause the formation of sand waves at the sediment-water interface. The depth of the floodwater becomes less and less after a considerable thickness of the cross-bedded sand was laid down on the point bar. Meanwhile, the speed may have remained the same, or even quickened.

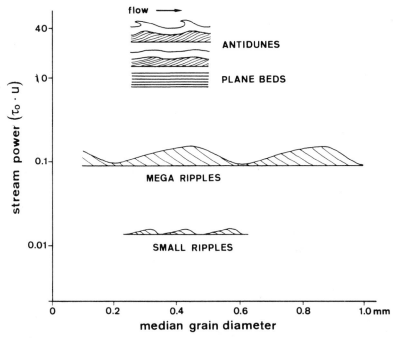

Fig. 7.4. Bedform and stream power (after Allen 1970). With an increase of stream power, the bedform of a stream channel or flume underlain by unconsolidated sediments changes from small ripples to mega-ripples, to plane beds to antidunes

The velocity/depth ratio increases to such an extent that sand waves disappear and horizontally laminated sand was deposited on a flat surface. Eventually, the flood slackens, the current velocity decreases drastically. At the same time the water depth becomes less, but the velocity decrease renders again the Froude number so small that a rippled bedform appears. This was the condition for the deposition of cross-laminated silts. Finally, the mud settles out of the suspension of a standing water-body after the flood recedes.

Not all alluvial deposits are characterized by a point-bar sequence. Braided stream deposits more commonly show laminations parallel to the channel or bank surface than cross-bedding. Where the channel surface is horizontal, these sediments are horizontally laminated or bedded. Where the channel, or the sand or gravel bank, has a gentle inclination of 5° or 10°, the bedding surface parallel to the channel has correspondingly a 5° or 10° inclination. Such a structure of inclined bedding is not called cross-bedding, but *accretional bedding*.

The common presence of horizontally lamination or accretional bedding in sands deposited by a braided stream is evidence that the bedform of braided channels is commonly a flat bed, rather than sand waves. The horizontally laminated structure indicates a relatively high velocity/depth ratio. Floods spreading out in braided channels have commonly a Froude number too large to make sand waves at the sediment-water interface.

Why should sand waves form at all in deeper or slowly moving rivers? Why should they disappear when depth decreases or current speed increases?

I have not found any profound theoretical explanation. R. A. Bagnold, one of the greatest theoreticians on sand waves, suggested that a stream makes its own autobahn.

As Chezy's equation shows, a stream has to flow as fast as the gradient and depth dictate. When it begins to accelerate, either the gradient, or the depth has to be decreased or the resistance be increased to reduce the acceleration.

Bagnold speculated that Nature tends to conserve its energy by striving for equilibrium. As we discussed in Chapter 3, meandering is a form of equilibration, seeking to decrease the flow speed by reducing the gradient. Braided channels provide another means of equilibration: acceleration is minimized by a decrease of depth and an increase of resistance. The changing bedforms may be still another way for a stream to seek its equilibrium. Where a stream channel is too deep or a gradient too steep, the acceleration of the fluid flow could be minimized or avoided if the resistance is increased through the formation of sand waves on the channel floor. Where it has to run faster, a stream eliminates sand waves to minimize resistance.

Perhaps the contrast is best understood if we understand the meaning of independent and dependent variables in a relation: When we drive an automobile, we slow down when we encounter a bumpy road, and speed up when we get on a new autobahn; the road surface is the independent variable and the driving speed has to be adjusted to fit the "bedform". A stream, on the other hand, can adjust its bedform to fit its speed: Where there is a "speed limit", a bumpy road is made to reduce the driving speed. Where hurrying is necessary, a stream makes its own autobahn as it flows.

Small and large sand waves and a flat bed are bedforms at the sediment-water interface, when there is little wave motion at the water-air interface. Flume experiments with uniform flow depth but increasing flow speed show the progressive change from ripples to dunes to flat bed (Fig.7.4). When the fluid flows at still greater speed, large waves propagating upstream, begin to appear on the water surface (Fig. 7.5). At the same time, sand waves are formed at the sediment-water interface, and their motion is in phase with that of the surface waves. As the Froude number of the fluid flow approaches unity, the sand waves and the surface waves both increase in amplitude, until the wave front becomes so steep that the waves break

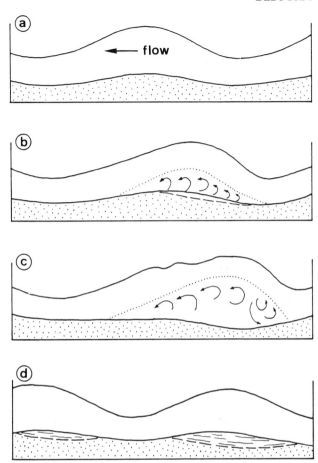

Fig. 7.5. Antidunes (after Middleton and Southard 1978). Antidunes form as a consequence of the breaking(**b**) of a standing wave (**a**) to erode the bottom (**c**). The faintly laminated antidune cross-bedding (**d**) dips in the upstream direction

abruptly, throwing much sediment into suspension. The sand waves migrate thus upstream in a direction opposite to that of normal dune migration. This type of bedform has therefore been called *antidunes*.

While the critical Reynolds number (**Re** ≈ 2000 for wide open channels) is the criterion to distinguish laminar from turbulent flows, the critical Froude number (**Fr** ≈ 1) distinguishes supercritical from subcritical flows.

Four flow regimes in an open channel are possible. As shown in Fig. 7.6, streams a few meters deep flowing at speeds less than 1 m/s are mostly turbulent, subcritical flows. Therefore, cross-bedding and horizontal lamination are the two most common types of internal structures in river sands.Turbulent supercritical flows are very fast flows in very shallow channels and are seldom seen in nature. In my many years of experience, I have seen antidune cross-bedding only twice: Once in an alluvial fan deposit of the Mojave Desert, another time in a braided-stream deposit of Tibet. Only the rare combination of a very steep gradient and a very wide channel has produced the speed/depth ratio, or the Froude number, necessary for sediment transport by the antidune type of fluid flow.

Dune migration is a very common sedimentary process, and cross-bedding a very common sedimentary structure. We see sand dunes in windy deserts, we encounter sand waves on riverbanks and in tidal channels, and we recognize sand waves in seismic profiling records of shallow sea bottoms. The presence or absence of cross-bedding is rarely a sufficient criterion for interpreting depositional invironments; we have to know what other sedimentary structures are present. The concept of a facies model postulates that certain associations of sediments, each with its characteristic structure, could be the unique deposit of one particular depositional environment.The flooding of the Brazos, and only this kind of flooding of a river, could bring about such an association as the point-bar sequence, so that this sequence can be illustrated in a picture-book to typify the deposits of meandering rivers. Are there other equally illuminating facies models?

Field geologists working in the Swiss Alps have long recognized two vastly different types of sandstone formations: the Flysch and the Molasse. The Flysch are Cretaceous and Early Tertiary formations characterized by the graded bedding of sandstone, whereas the Molasse are mid-Tertiary formations characterized by an abundance of thick, coarse conglomerate beds and the not uncommon presence of cross-bedded sandstone. British scientists working on their island have also observed two types of sandstones: those with graded bedding deposited in ancient *geosynclines* and those with cross-bedding deposited on ancient marine shelves (Fig. 7.7).

We have stated that cross-bedding is the internal structure of sand waves which are produced at the sediment-water interface of alluvial channels, of marine shelves, and at the sediment-air interface of the desert. Sand with cross-bedding is associated with sediments having other structures to make a facies model to characterize a point-bar sequence of a meandering-river deposit. Graded bedding, as we have discussed, is a signature of deposition from a turbidity current. What kind of facies model includes deposits typified by graded beds?

A graded bed in a flysch formation is, as a rule, less than 1 m thick. A closer inspection of such a sandstone layer will reveal other internal

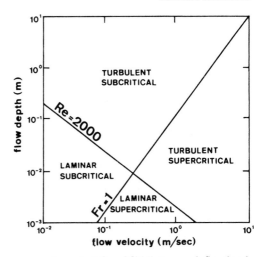

Fig. 7.6. Flow regimes in an open channel (after Middleton and Southard 1978). The criterion of turbulence is the Reynolds number and the criterion for transition to the supercritical regime is the Froude number of a fluid flow

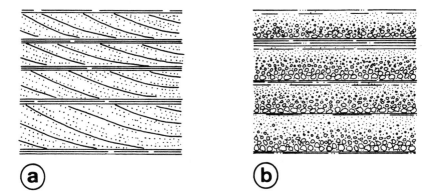

Fig. 7.7. Cross-bedded (a) and graded-bedded (b) sandstones

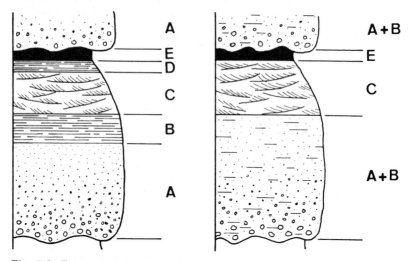

Fig. 7.8. Facies models of turbidity-current depositions. *Right side:* Bouma Model, *left side:* proposed model (see text for explanations)

structures. Where the sand contains silty material, faint horizontal laminations could be visible (Fig. 7.8). The structure suggests that the sand particles were laid down on a flat bed after settling from suspension. Not uncommonly, the graded bed is topped by a cross-laminated silt with no distinct layering between the two. The set or sets of cross-laminated silt could be several centimeters thick. In places, the cross-laminations have been deformed, apparently by the shearing force of the suspension flow above, to form *convolute bedding* .

The deposition of cross laminated silts indicates a different bedform; the bottom was rippled during the waning phase of the turbidite deposition.

Such internal structures of graded beds have been observed in Ordovician turbidites of the Appalachian Mountains by geologists of the New York State Geological Survey, but these observations, made during the 1920s, were ignored because no theoretical explanation could be given. When Arnold Bouma, a student of Ph. Kuenen, described the structures in the flysch sandstones of the French Alps, studies on flow regimes by scientists of the U. S. Geological Survey were carried out. Bouma was able to relate the structures to the changing flow regimes.

The typical Bouma sequence is divided into five divisions; they are from top to bottom (Fig. 7.8):

e) mud;

d) horizontally lamilated silt;

c) cross-laminated silt;

b) horizontally laminated sand (or coarse silt);

a) graded bedding (sand to coarse silt).

The Bouma sequence is beloved by those who like to use the picture-book approach to geology. They might state that a certain sedimentary bed is a turbidite because it exhibits an incomplete Bouma sequence, even if the bed in question is a cross-laminated silt deposited by a marine bottom current.

I never liked Bouma's scheme for the following reasons:

1. The classification is based upon three sets of criteria: texture, structure, and lithology, so that the divisions are not mutually exclusive.

2. Division *e,* mud or clay, may or may not be part of the turbidite, because much of the clay particles may have been deposited as a hemipelagic sediment after the turbidite deposition.

3. Division *d ,* horizontally laminated fine silt, is more an exception than a rule.

4. Division *c*, cross-laminated silt, as we will discuss, may or may not be deposited by turbidity current.

5. Division *b*, horizontally laminated sand can be observed at the very bottom of a graded bed.

6. Division *a*, graded bedding is not restricted to the lowest part of a turbidite, but is the characteristic of the whole turbidite bed, although the grading is commonly more obvious in the lowest part where the grain-size change is readily discernible.

Using bedforms and corresponding flow regimes as a criterion for classification, a turbidite can only be divided into two parts: the horizontally (division *b*) and the cross-laminated (division *c*).

If we exclude the lag gravels in a point-bar sequence, the deposit of a spring flood on a point bar consists of three divisions:

C) cross-laminated (silt);

B) horizontally laminated (sand);

A) cross-bedded sand.

The two divisions of a turbidite, classified according to the bedforms, are:

C) cross-laminated (silt);

B) horizontally laminated (sand or silt).

The division A (cross-bedded sand) is almost always missing in a turbidite bed. Is there a physical basis to support the empirical observation that turbidites are almost never cross-bedded? There could be an answer if we apply the experimental observations of bedform in flume studies to interpret the sedimentary structures of turbidites. The observations indicate that the sand waves appear at the sediment-water interface when the Froude number is smaller than a certain critical number, say 0.35 or 0.40. What is the Froude number of a turbidity current?

If the Froude number is taken as an expression of the ratio of the inertial force to gravity force in a turbidity current, we have for the *densiometric* Froude number

$$\mathbf{Fr_d} = \frac{u}{\sqrt{\frac{\Delta \rho_c}{\rho} g \, D}} \quad .$$

The effective density of a turbidity current $\Delta \rho_c$ is the difference between the density of a current and that of seawater, and is commonly at least two orders of magnitude less than unity. Consequently, for currents flowing fast enough to transport sand by suspension, the densiometric Froude number is almost always larger than the critical number marking the transition of bedform from sand waves to flat bed. The absence of cross-bedding in turbidites is, therefore, to be expected from a consideration of the flow regimes.

In fact, it is surprising that antidune structures are not more common in turbidites in view of the commonly small values of $\Delta \rho_c$. I have only seen one turbidite, in an Oligocene formation on Barbados, which exhibited antidune structures. Small values of the densiometric Froude number may be an indication of the relatively great thickness of marine suspension currents. The turbidity current generated by the Grand Banks sediment-gravity flow had a speed exceeding 10 m/s. For such a current not to deposit antidune sand, the densiometric Froude number must have been less than unity. Thus the current thickness must have been more than 100 m if $\Delta \rho$ is 0.1 g/cm^3. Such numerical estimates are, incidentally, about the same order of magnitude as those at Cable K or Cable L, calculated by Kuenen and by Menard on the basis of Chezy's equation and those by Lüthi on the basis of assuming auto-suspension.

Suggested Reading

Geologists engaged in model studies should read M. King Hubbert's *Theory of scale models as applied to studies of geologic structures* (Geol Soc of Am Bull 48:1459-1520, 1937) and Langhaar's book on *Dimensional analysis and theory of models*, 1951, cited in Chapter 3.

Middleton and Southard, in their book on *Mechanics of sediment movement*, 1978, cited in Chapter 3, reviewed the relation between flow regimes and bedforms. The pioneering experimental work on the subject by D.B.Simons and E.V.Richardson has been reported in numerous publications; their article (with C.F.Nordin) on *Sedimentary structures generated by flow in alluvial channels* is published in a journal easily accessible to a geologist, namely, Society of Economic Paleontologists and Mineralogists(Spec Publ 12:34-53, 1965). The subject has also been treated by John Allen (1970) in his book cited in Chapter 3.

The distinction between cross-bedded and graded-bedded sandstones was appreciated by E. B. Bailey as early as 1936 (*Sedimentation in relation to tectonics*, Geol Soc Am Bull 47:1713-1726). Hugh Bernard and his associates proposed their point-bar model in *Recent sediments of southern Texas*, Guidebook 11 published in 1970 by the Bureau of Economic Geology of Texas. Bouma presented his facies model for turbidites in his 1962 book *Sedimentology of some fysch deposits* (Amsterdam: Elsevier).

8   Oceans Are Ventilated

Ocean Currents- Bernoulli's Theorem - Darcy-Weisbach Equation
Contourites

We have followed the fate of a sedimentary particle from mountains to the mouth of a river where floodwater is spread out as a plume. What happens to it afterwards?

Sedimentary particles released from the plume after its mixing with ambient seawater would either sink or be swept into ocean currents and transported farther before they finally come to rest.

The time elapsed for a particle to settle out of a surface current is the ratio of the depth of ocean current d and the settling velocity u_S. During this time, the particle has traveled with the current at a speed u over a distance s. Equating $t = d / u_S = s / u$, we have

$$s = \frac{u}{u_s} d \quad .$$

$$(8.1)$$

Sand and silt particles with large u_S cannot be transported far away from the delta and are mainly deposited as deltaic or coastal sediments. Clay-sized particles, with a settling velocity less than 1.6×10^{-3} cm/s, suspended in a current of 300-m depth and 50 cm/s velocity, could be carried halfway around the world. Some surface currents, such as the Florida Stream, have higher speed and greater depth; it is thus not surprising that *pelagic* clays are the normal terrigenous sediments on the bottom of the open ocean. The term pelagic is derived from the Greek word for swimming; pelagic sediments literally swim their way into open oceans to their final resting place.

Detrital sediments on a continental slope and continental rise include considerable silt, and, in places, sandy components. They are *hemipelagic,* meaning not entirely pelagic; some of the particles did not swim in surface currents, but were dragged to their site of deposition by bottom currents. What are these bottom currents? How do they originate?

In the early 1950s, when the idea of turbidity-current deposition was sweeping across the academic community, there was the tendency to call all deep-sea sandy sediments turbidites. I was working in the Ventura Basin, then, where the thick-graded sands showed many features characteristic of

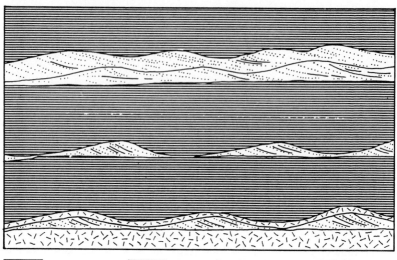

Fig. 8.1. Cross-laminated silt of Ventura Basin (after Hsü 1964). Turbidite layers are thin blankets of sediment, but the deep marine cross-laminated silt of Ventura has a patchy distribution. The postulate that they have been deposited by a bottom current is confirmed by the fact that currents, depositing cross-laminated silts, have been measured on modern seafloor

turbidite deposition. The top of the turbidite sands is commonly cross-laminated, forming the so-called division C of the Bouma sequence. Such cross-laminated sediments could indeed be deposited by turbidity currents, as our flume studies confirmed. However, I was puzzled by the fact that many thin siltstone beds of Ventura showed no other signatures of turbidite-deposition than cross-lamination (Fig. 8.1).

A common explanation, then and now, is that cross-laminated silts are deposited in the waning phase of the turbidity-current transport and distant from origin as *distal* turbidites. I could go along with the first statement, but the term distal is misleading because cross-laminated silts are common in the Pliocene submarine canyons and deep-sea fans of the Ventura Basin. These sediments must have been deposited at sites proximal to, not distal from, their origin up the canyon. Distal deposits of weak currents are laid down where proximal deposits of strong currents are found. This lesson was enough for me to shy away from such a picture-book approach in sedimentology, and from the use of misleading terms such as distal and proximal.

Numerous features characteristic of the Ventura cross-laminated silts led me to consider the role of currents other than turbidity currents in deep-sea sand deposition. The cross-laminated layers consist mainly of well-sorted coarse silt. They are present locally as discontinuous, patchy deposits between thin layers of shale. Unlike turbidites, which are characterized by transported microfaunas, the cross-laminated silts of Ventura have a deep marine benthic fauna, not much different from that in the overlying and underlying shale beds. The foraminiferas had apparently lived and died on silty bottom as the cross-laminated sediments were being deposited.

Finally, on a hot summer afternoon, I fought my way through the sagebrush to find a well exposed canyon wall. I saw cross-laminated silt between two thin laminae of ashfall. This was one last bit of evidence which convinced me that some cross-laminated silts at least are not turbidite: Newly deposited volcanic ash with its high water content could be easily stirred up by a turbidity current fast enough to carry coarse silt particles. Yet I found little erosion of the ashfall. Also the silt is clean, i.e., well sorted, almost free of clay-sized particles. It is different from a silt dumped out of a suspension, be it a waning flood, or the waning phase of a turbidity-current transport. The patchiness of the rippled silt also impressed me: A turbidity current, no matter how weak, always deposits a continuous layer of sediment, however thin it might be. If there had been bottom currents capable of producing a rippled bedform, coarse silt or very fine sand could be concentrated on rippled crests thus forming a discontinuous patch of cross-laminated sediment, if the sediment supply is insufficient.

Already in the early 1950s, deep-sea photography indicated the presence of a rippled deep-sea bottom and current meters detected the presence of bottom currents flowing down submarine canyons off the coast of California or flowing parallel to the continental margin of the eastern Atlantic. The speed of the currents ranges between 15 to 30 cm/s, fast enough to erode sandy or silty bottom. I came to the conclusion, therefore, that cross-laminated sediments in graded-bed sequences are not all turbidites. The cross-laminated silt of Ventura, for example, could be a bottom-current deposit. Of course, the sand and silt particles did not all come directly from the shelf areas. Much of the course debris may have been brought down by turbidity currents and deposited as turbidites. The finer sand and silt particles in these are eventually eroded and transported by bottom currents and redeposited as cross-laminated sediments (Fig. 8.2). This interpretation was considered "counterrevolutionary" in the days when the "revolution of the turbidophiles", namely, those radical advocates of turbidity-current deposition, was reaching its zenith of popularity.

Over the last 20 years, we have acquired much understanding about bottom currents from the work of oceanographers and marine geologists.

There is the vigorous Mediterranean Undercurrent, carrying a "hot salty tongue" of water through the Strait of Gibraltar into the Atlantic. This hot salty tongue has originated from the eastern Mediterranean where the surface

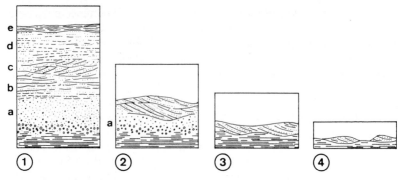

Fig. 8.2. Genesis of cross-laminated silt through the reworking of a turbidite. **1** Deposition of a turbidite bed by turbidity current; **2** removal of turbidite divisions *e, d, c, b* , and the deposition of cross-laminated silt on a graded bed; **3** deposition of cross-laminated silt on hemipelagic seafloor; **4** disrupted layering where sediment supply is insufficient

water has almost 39°/oo salt content because of evaporation. The "hot brine" flows westward as an intermediate water mass and descends to about 400 m depth near the Strait of Oranto, where it meets the colder, intermediate water mass from the Adriatic (Fig. 8.3). Part of the more salty water is cooled and descends to form the bottom circulation of the eastern Mediterranean Sea. Another part continues westward and joins the evaporated water originating from the south of France to form the "hot salty tongue" which eventually spills over the sill of the Strait of Gibraltar. The flow west of the strait descends to a depth of about 1200 m. The sea bottom under the current, commonly muddy but locally sandy, is smooth or rippled; current velocities of more than 20 cm/s have been measured. At distances greater than 250 km, the current leaves the bottom to become the Mediterranean Intermediate Water Mass of the Atlantic Ocean.

There is also the famous Western Boundary Undercurrent on the western boundary of the Atlantic Ocean. Oceanographers have now traced its origin to the Antarctic, where a warm intermediate water mass from a lower latitude rises. This water is chilled to subzero temperature. With sea ice forming near the Weddell Sea, what is left over becomes even saltier, and this "cold salty tongue" of water then descends to form the Antarctic Bottom Water Mass (AABW), which flows northward as the Western Boundary

Fig. 8.3. Surface and intermediate Mediterranean circulations (after Lacombe and Tchernia 1971). **a** Surface current. **b** Intermediate current. **c** Bottom circulation

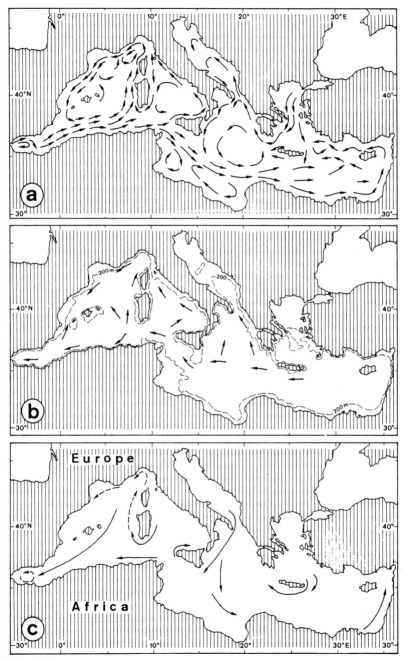

Fig. 8.3

Current. After reaching the North Atlantic Ocean, the bottom current makes a right turn and is joined by another salty tongue from the Arctic known as the North Atlantic Deep Water Mass (NADW), which has been spilled over the sills of the Norwegian Sea. The combined bottom current then flows southward and penetrates almost as far as the Antarctic Polar Front, before turning east and flowing into the Indian Ocean, and ending up in the North Pacific.

Studying the sediment cores of ocean drilling, paleoceanographers have found that this system of Atlantic bottom circulation could be traced back to the Late Eocene or Early Oligocene epoch some 40 million years ago. The current is not only strong enough to transport sediment particles for long distances along its path, it also has erosive power. On both margins of the Atlantic, *unconformities*, or large gaps in the sedimentary history, are present in sedimentary sequences of deep sea origin; the discovery has dispelled the traditional prejudice that erosion is exclusively a subaerial process. The initiation of AABW took place at about the same time when the ice cap at the South Pole expanded. As I have indicated, the origin of AABW is related to the formation of sea ice; icy-cold seawater rendered more salty, and therefore denser, by freezing sinks to form this bottom current.

What is the physical basis of the current circulation of the oceans?

Surface currents are generated by the shearing of wind stress, and winds are movement of air masses because of pressure differences. Bottom currents move due to pressure differences in seawater. The movements of the water masses of the oceans are thus primarily driven by pressure forces; tidal and coriolis are secondary factors, which serve to influence the circulation patterns of ocean currents.

The distribution of pelagic sediments in the oceans is related to the circulation of surface currents. Wind-generated upwelling is particularly important for sedimentary processes: On the west side of continents, for example, strong winds from onshore blow away the surface water, which is replaced by cooler waters from the depth. The deeper water has its ultimate origin from polar regions. On the west side of South America, for example, the cold water of the Antarctic is carried north by the Humboldt Current. The upwelling of this nutrient-rich water promotes plankton blooms. Consequently, phosphate deposits and biogenic sediments, especially those consisting of siliceous skeletons, are very common in this region of upwelling.

Wind-generated stress also causes the upwelling of the nutrient-rich water mass of equatorial currents. Radiolarian oozes, which consist mainly of the siliceous skeletons of this zooplankton, have been deposited in tropical oceans since the Paleozoic Era. The close interrelation between nutrient supply and paleogeography permits the reconstruction of paleoceanographic circulations.

The wind-generated stress can, however, only penetrate to a limited depth beneath the ocean surface. All surface currents have thus a limited depth, commonly in the order of several hundred meters. The boundary of a current is manifested by steep gradients in salinity and in temperature between the current and the enclosing ambient seawater.

Bottom currents are not wind-driven; they originate because denser water descends. The driving force is gravity in the form of a pressure head. What is a pressure head?

My aunt built a new house when I was 8 years old. She had a spraying fountain in her garden. When I visited her during my vacation, I always asked her as a favor to let me turn on the faucet. My fascination is understandable: Children are used to seeing objects fall, but why should a jet of water shoot skyward; it seemed to defy gravity.

Water in rivers, or in open channels, flows downslope. Water in pipes can, however, flow up as well as down slope; Chinese farmers in mountainous regions knew that and, for thousands of years, they have been using bamboo stems to pipe water for irrigation. Water in pipes can move upward under pressure, like the water jet spraying out of my aunt's fountain.

Yet the upward movement is limited. I remember one incident on a visit to Bulgaria. At Plodiv, after a hot day of field excursion, I was given a room on the top floor, where the view was the best; our modest hosts had rooms of moderate price somewhere lower down. When we met for dinner, I noted that they had all changed, while I remained in my field clothes because I had no water for a bath.

"Do you have hot water?" I asked.

"Yes, of course. Why, don't you?"

No, I did not have any water, because it was not able to reach the tenth floor, but they had plenty down there on the first floor.

We all know that the ability of water to go uphill is limited, depending upon the pressure head. What is the relation of pressure, flow rate, and plumbing?

We could start from the fundamental equation in hydrodynamics, the Bernoulli Theorem, which states

$$\frac{p}{\rho} + g\,H + \frac{u}{2} = \text{constant} , \qquad (8.2)$$

where p is the pressure, ρ the density, u the speed, H the height of the hydraulic head, and g the gravitational acceleration.

This theorem, formulated by Daniel Bernoulli in 1738 on the basis of Newton's Second Law of Motion, can be considered a statement of the conservation of energy. The three terms in Eq. (8.2) denote three forms of energy per unit mass. When you multiply those terms by mass, m or $\rho\,V$, you get:

$$p V + m g H + \rho V u^2 / 2 = \text{constant} . \tag{8.3}$$

The first term is the work done by pressure, the second is the potential energy, and the third kinetic energy.

Consider now the case of the horizontal flow in a pipe past an obstacle (Fig. 8.4), the conservation of energy, as expressed by the Bernoulli Theorem, states that

$$\frac{p_1}{\rho} + 0 = \frac{p_0}{\rho} + \frac{u^2}{2} \tag{8.4}$$

or

$$\Delta p = p_1 - p_0 = \rho \frac{u^2}{2} . \tag{8.5}$$

Such a pressure difference has been called *stagnation pressure*. This is the pressure you feel when you stick your hand out of a car window when it drives speedily on a freeway. The faster the driving speed, the greater the pressure. This is, in fact, the principle which enables us to measure the speed of fluid flow indirectly through the measurement of the stagnation pressure.

Integrating over the cross-sectional area (A) of the obstacle, the pressure force is

$$F_p = \iint p \, d A . \tag{8.6}$$

All forces have the same dimensions. Although the resistance force of water flow in a pipe is a shearing resistance, or a different kind of force, it has to have the same dimension and can thus be expressed in terms of the pressure force (Eq.8.5). The two can be related by a dimensionless number, the resistance coefficient C_f

$$F_r = C_f \frac{\rho u^2}{2} A . \tag{8.7}$$

where A, in this case, is the contact area between the pipe and the flow and is equal to $\pi D L$, where D is the diameter of the pipe and L the length of the pipe segment. Note the similarity of this equation to Eq. (3.7) relating resistance to fluid flow in an open channel. We have, therefore, also:

$$F_r = \tau_r \pi D L ,$$

where τ_r is the shearing stress exerted by the fluid on to the inner wall of the pipes.

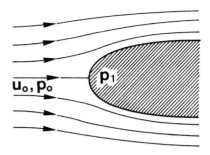

Fig. 8.4. Stagnation pressure. Current flowing with speed u_o and p_o will exert a pressure p_1 at the point of stagnation

The work done by the pressure force driving the flow of water in a pipe has been converted from potential energy. If the energy loss due to friction is negligible, the kinetic energy of the fluid could be converted back into potential energy. In that ideal situation, water should have climbed up to the 10th floor of the Bulgarian hotel where I stayed, as the 10th floor was not higher than the water tower of the city. In fact, friction is not negligible. The friction loss, over a horizontal transport distance ΔL, is equal to the loss of the head ΔH, or

$$F_r \Delta L = \rho V g \Delta H . \tag{8.8}$$

For water flow at constant velocity, the resisting and driving forces are equal and opposite, the pressure force is thus

$$F_p = F_r = \rho V g (\Delta H / \Delta L) . \tag{8.9}$$

We encountered the expression $\Delta H / \Delta L$, or dH / dL, during our discussion of sediment-gravity flows. It was called the energy line, or the energy gradient. In the case of water flow in pipes, the ratio has been referred to as the gradient of the head loss, or the hydraulic gradient.

In a steady state flow the shearing force $\tau_r A$ is equal to the pressure force driving the movement of water in pipes. We have

$$\tau_r \pi D L = \rho \, \pi \left(\frac{D}{2}\right)^2 L \, g \left(\frac{\Delta H}{\Delta L}\right)$$

or

$$\tau_r = \rho \, g \left(\frac{D}{4}\right)\left(\frac{\Delta H}{\Delta L}\right) . \tag{8.10}$$

The loss of the hydraulic head is caused by friction, and the frictional loss of energy can only be evaluated by experimentation. The loss is always proportional to the travel distance ΔL; it is more for greater flow velocity, but is less for pipes of greater diameter. The experimental relation is

$$\Delta H = f \cdot \left(\frac{\Delta L}{D}\right) \cdot \frac{u^2}{2g} \quad ;$$

$$\left(\frac{\Delta H}{\Delta L}\right) = f \cdot \left(\frac{1}{D}\right) \cdot \frac{u^2}{2g} \quad ; \qquad (8.11)$$

where f is an empirical number determined by experiments, and is called *friction factor*, or Darcy-Weisbach coefficient.

The term $(\Delta H / \Delta L)$ is dimensionless, the ratio $(u^2/ g D)$ is the square of the Froude number and thus also dimensionless. The two are related by the dimensionless Darcy-Weisbach coefficient.

Substitute Eq. (8.11) into Eq. (8.10), we have

$$\tau_r = \rho\, g\!\left(\frac{D}{4}\right) \cdot f \cdot \left(\frac{1}{D}\right) \cdot \frac{u^2}{2g} = \rho \cdot \frac{f}{8} \cdot u^2$$

$$\tau_r = \frac{f}{4} \cdot \frac{\rho u^2}{2} \quad . \qquad (8.12)$$

This is the so-called Darcy-Weisbach equation relating steady-state resisting stress to the square of the flow velocity. A comparison of Eqs. (8.7) and (8.12) indicates

$$f = 4\, C_f \ , \qquad (8.13)$$

where the dimensionless numbers expressing frictional resistance in the Nikuradse curves or the Moody diagram are the friction factor f. Equation (8.13) is the basis of my previous statements that the value of the resistance coefficient C_f is one-fourth that of f, and that we could use the readouts from the Nikuradse curves to compute fluid resistance involving C_f.

We can now solve for flow velocity u from Eq. (8.11), thus

$$u = \sqrt{\frac{2g}{f} \cdot D \cdot \frac{\Delta H}{\Delta L}} \quad . \qquad (8.14)$$

In other words, the speed of pipe flow is related to the diameter of the conduit and to the hydraulic gradient. Note that this equation is identical to Chezy's equation (Eq. 4.1), if the slope of the head ($\Delta H/\Delta L$) loss is substituted by the slope of the channel, s, and the diameter of pipe, D, is substituted by the depth of channel, d., and the friction factor by the resistance coefficient.

The ocean bottom currents flow under pressure like water in pipes. The pressure difference is, however, not derived from the height difference, but by the density difference between the fluid flow and the ambient water mass, or according to Eq.(8.9)

$$F_p = \Delta \rho \ V \ g \ (H / L) , \qquad (8.15)$$

where H is the ocean depth in the path of bottom-current flow, and L is the horizontal distance of flow. The greater the density contrast, the greater the force, the greater is the strength of flow. The density contrast can also be expressed in terms of pressure or

$$\Delta \rho \ g \ H = \Delta p . \qquad (8.16)$$

We have therefore

$$F_p = \Delta p \ A \ . \qquad (8.17)$$

In our discussions of the bottom currents, we have mentioned the "hot salty tongue" from the Mediterranean and the "cold salty tongues" from the polar regions. The bottom circulation generated by the first type is called halokinetic, and that by the second type is called thermohaline circulation. Halokinetic circulations may have been important during the Mesozoic, when polar ice caps were small or non-existent. Thermohaline circulation has become dominant since Late Eocene, when the Antarctic Ice Cap began to expand in the region of the South Pole. In both cases the pressure drive originates from a density difference as shown by Eq.(8.16).

The Western Boundary Current flows northwards, parallel, more or less, to the contours of the continental slope. Sediments deposited by this type of bottom current have been called *contourites*. The cross-laminated sands and silts of the Ventura Basin include contourites as well as turbidites. This brings me back to the discussion of the so-called Bouma sequence. It has been stated that the Bouma sequence is "incomplete in many turbidites; in numerous instances, only the division C is visible." Such a statement could be incorrect. Where cross-lamination (Bouma's division C) is the only visible sedimentary structure, we do not have a facies model for unique interpretation and we would not be in a position to differentiate a turbidite from a contourite.

Sediments redeposited by bottom currents are not necessarily silt or sand.

Remarkable ripple marks have been photographed in abyssal brown clay of the Puerto Rico Trench beneath the Antarctic Bottom Water Mass. Redeposition of sandy sediments by bottom currents takes place, as a rule, only in regions where coarse detrital sediments have been brought down by turbidity currents. The bottom sediments in the path of bottom currents can, however, also be muds, clays, or manganese nodules.

The movement of bottom-current circulation serves not only to rework previously deposited sediments, the circulation is essential for the oxygenation of deep-sea environments. The water of the Black Sea is devoid of oxygen at a depth greater than 200 m. Its surface water is brackish, having been derived from the mixing of river water and salt water from the Mediterranean. This less dense surface water cannot sink so that there is no driving force for bottom circulation. The salt water, which found its way across the Bosporus into the Black Sea, stays beneath the thermocline till all its oxygen has been expended through the oxidation of organic remains. As a consequence, sediments rich in carbon or hydrocarbon are preserved in such an *euxinic* environment (pertaining to Mare Euxenus, which is the ancient name of the Black Sea). Such sediments are potential source beds for hydrocarbons.

Anoxic sediments of Neogene age are common in deep basins of the eastern Mediterranean Sea, deposited at a time when the density contrast between the evaporated eastern Mediterranean brines and the ambient seawater was not large enough to induce bottom circulation.

Anoxic sediments are very rare in Cenozoic sequences of the open oceans. Yet black sediments rich in organic carbon are very common in sedimentary formations of Cretaceous age in the Atlantic Ocean. They constitute important source beds for the petroleum occurrences on both sides of the Atlantic. Apparently, the driving force for bottom circulation was relatively weak prior to the Tertiary glaciation of the polar regions.

Suggested Reading

The derivations for the Bernoulli Theorem and for the Darcy-Weisbach equation can be found in hydrodynamics textbooks; one may look up Prandtl, for example, cited in Chapter 3. The principles of ocean circulations are discussed in oceanography textbooks such as the one by Grant Gross (Englewood Cliffs, New York: Prentice Hall, 1977). We have learned a great deal about oceanic, especially abyssal, circulations during the last few decades. For those who are interested in finding out how we used to grope in the dark, they might look up the articles by Walter Munk on *The circulation of the oceans,* and by Henry Stommel, *The circulation of the abyss,* published by Scientific American in 1955 and 1958, respectively. A brief review on the Mediterranean circulation by H. Lacombe and P. Tchernia is included in the book *The Mediterranean Sea* (Stroudsboug, Pa: Dowden, Hutchinson and Ross, pp 25-36, 1971).

Sometime during the 1950s, I first became aware of the importance of bottom currents as a sedimentological agent, and my idea was published in an article on *Cross-laminations in graded bed sequences* (J Sediment Petrol 34:379-388, 1964). Meanwhile, Bruce Heezen showed me pictures of ripple-marked sand and silt deposits on modern deep-sea floor and he coined the term contourite. Some of the excellent photographs found their way into his book with C.D.Hollister on *The face of the deep* (New York: Oxford Univ Press, 1971). We have indeed come a long way since those days, and the current knowledge is summarized by K. Pickering and others in their review article on *Deep water facies, processes and models* (Earth Sci Rev Vol 23, 1986).

9 Groundwater Circulates

Darcy Equation - Poiseuille Law - Hydrodynamic Potential
Permeability - Compaction - Diffusion

I joined the Shell Development Company in 1954 to conduct scientific research in order to improve the technology of petroleum exploration and production. After a year working on the Gulf Coast, my second assignment was to investigate the sandstone reservoirs of the Ventura Avenue Field, California.

The Ventura Basin is situated northwest of Los Angeles. The basin started to subside rapidly in Late Miocene and reached its greatest depth in Early Pliocene. A Pliocene/Pleistocene sedimentary sequence, more than 5000 m thick, had been deposited before the sequence was folded during the Middle Pleistocene. Faunal ecology indicates Miocene and Pliocene deposition in the deep sea, before the basin was largely filled up in Early Pleistocene when shallow marine clastics were laid down.

The resedimented deposits of the basin include four facies: (1) hemipelagic mud, the normal sediments of the basin; (2) conglomerate with mud matrix, laid down by sediment-gravity flows in submarine canyons and fans; (3) graded sandstone, transported to, and deposited in basin troughs by turbidity currents; and (4) cross-laminated siltstone, turbiditic detritus reworked by bottom currents (Fig. 9.1).

The Ventura Avenue Field was discovered during the 1920s. Petroleum has been trapped in a faulted anticline, and the reservoir beds are turbidite sandstone (Fig. 9.2). Ventura was one of the major oil fields of California, and its production reached the high point during the early 1950s, when I joined Shell. With the shallower producing zones exploited, the development wells had to be drilled deeper and deeper, penetrating finally, Upper Miocene sediments. It was discovered that the reservoir sands are less permeable at depth, although commercial production was still possible. I was then asked to investigate this unfavorable permeability trend and to evaluate the further potential of development.

What is permeability? Permeability is a measure of the ease with which a fluid flows through a porous medium. This problem was first investigated by Darcy, a French engineer of the last century, who performed a famous experiment to evaluate the various factors influencing the rate of groundwater movement. Darcy's experiment was simple and elegant: Water seeps across a cylinder of length L filled with sand, and the rate of the flow

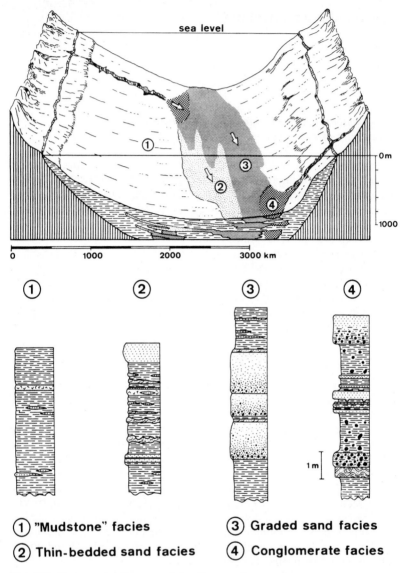

① "Mudstone" facies   ③ Graded sand facies

② Thin-bedded sand facies   ④ Conglomerate facies

Fig. 9.1. Deep-sea sediments of the Ventura Basin (after Hsü 1979)

Fig. 9.2. Ventura anticline (after Hsü 1977). Structural cross-section of Ventura field. *A, B, C, D,* and *N* are producing blocks. Note that the deepest producing formations lie more than 3000 m below the surface, where sandstone permeability is much reduced by compaction

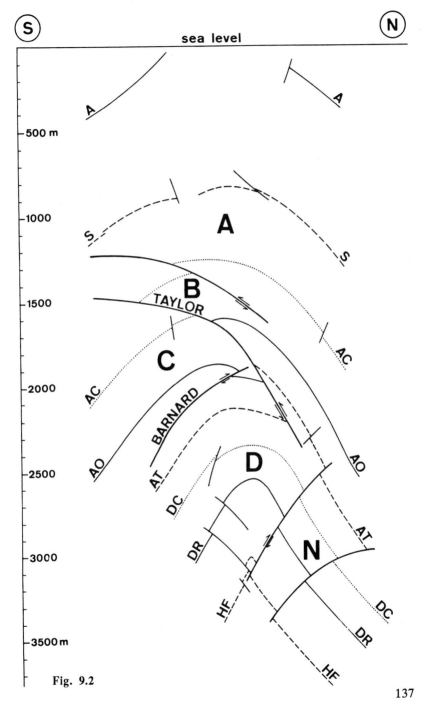

Fig. 9.2

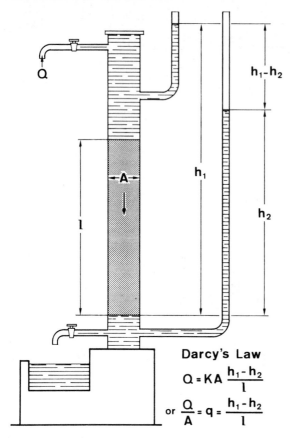

Darcy's Law

$$Q = K A \frac{h_1 - h_2}{l}$$

$$\text{or } \frac{Q}{A} = q = \frac{h_1 - h_2}{l}$$

Fig. 9.3. Henry Darcy's experiment on flow of water through sands (after Hubbert 1940). Darcy related the flow rate to the difference of manometer height. Darcy used mercury manometers to measure pressure. In this drawing, the pressure difference is expressed in terms of height difference of water manometers

volume (Q) is related to the height difference (ΔH) between the two ends of the cylinder, to the cross-sectional area A and to a quality K, which we now call transmissibility (Fig. 9.3), or

$$Q/A = K (\Delta H)/(\Delta L) . \tag{9.1}$$

Transmissibility is not dimensionless; the hydraulic gradient dH/dL is. The numerical number K has the same dimension as velocity, namely, centimeters per second. Given the same hydraulic gradient, the transmissibility of a porous medium is a function, *inter alias*, of viscosity; oil transmission is much slower than water flow.

Water flows through sand, because a network of interconnecting pores is present between sand grains. The motion of the water through a porous medium is driven by hydraulic pressure, like the flow of water through pipes. The driving force forcing the water through a tube of radius R is

$$F = (p_2 - p_1) \pi R^2 \ . \tag{9.2}$$

The fluid stress exerted by the flow is

$$\tau = \frac{F}{A} = \frac{(p_2 - p_1) \pi R^2}{2\pi r \Delta L} = \frac{\Delta p}{\Delta L} \frac{R}{2} \ , \tag{9.3}$$

where L is the length of the flow segment under consideration.

If the radius of the pipe is sufficiently small so that the resistance stress is the viscous resistance at the distance r from the wall of the capillary tube, we have

$$\tau_r = \eta \frac{d u}{d r} \ . \tag{9.4}$$

Equating the fluid stress and the viscous resistance and solving for flow velocity we have

$$u = \frac{\Delta p}{2\eta \Delta L} \int_r^R r \ dr = \frac{\Delta p}{4\eta \Delta L} \left(R^2 - r^2\right) \ , \tag{9.5}$$

where R is the inner diameter of the tube.

Equation (9.5) shows that the linear velocity of viscous flow through a pipe is not constant; the velocity depends upon the distance from the center of the pipe. The maximum velocity at the center is

$$u_{max} = \frac{R^2}{4\eta} \left(\frac{\Delta p}{\Delta L}\right) \ . \tag{9.6}$$

The velocity gradually decreases to null at the contact between the fluid and the inner wall of the pipe. A convenient measure of the rate of viscous flow is the rate of volume transport, or discharge

$$Q = A u$$

$$Q = \int_0^R 2\pi r \, dr \, u = \frac{\pi R^4}{4\eta} \left(\frac{\Delta p}{\Delta L}\right) \quad . \tag{9.7}$$

This empirical formula relating the discharge to the fourth power of the pipe radius, was derived independently by Hagen in 1839 and by Poiseuille in 1840, and has been called the Poiseuille's, or the Hagen-Poiseuille's law. The average linear velocity can be defined by

$$q = \frac{Q}{A} = Z R^2 \frac{1}{\eta} \left(\frac{\Delta p}{\Delta L}\right) , \tag{9.8}$$

where Z is a dimensionless number.

Groundwaters do not flow through a cylindrical tube, so that Poiseuille's law is not directly applicable. However, the interconnecting pores have an average diameter (D), and this diameter can be related to that of the tube by a dimensionless number, or

$$R = Z D \text{ , or:}$$

$$k = N D^2 , \tag{9.9}$$

where N is a dimensionless number and k is related to R^2 by a dimensionless number.

The concept of permeability is defined by Eq.(9.9), and it is a function related to the cross-sectional area of the pore space between grains in the rock. Substitute Eq. (9.9) into Eq. (9.8), we have

$$q = k \frac{1}{\eta} \left(\frac{\Delta p}{\Delta L}\right) . \tag{9.10}$$

Equation (9.10) is the formula, cited in numerous textbooks, relating the rate of groundwater flow to permeability and pressure gradient. M. King Hubbert, a noted geophysicist, who was my mentor when I first joined Shell, spent many years in calling attention to the professional community that this equation is, strictly speaking, not applicable to groundwater motion, because the derivation of Poiseuille's law has not considered the role of the body force of the fluid. In fact, Hubbert told me once that Eq. (9.10) can be easily proved wrong by a simple experiment:

We can construct a box and fill it with sand and water, with a sediment thickness ΔH. The pressure of the water at the top is the atmospheric pressure p_0 and that at the bottom is $p_0 + \rho g \, \Delta H$. There is a pressure difference between the top and bottom, and, according to Eq.(9.10), there should be flow from the bottom to the top of the box. Yet, we all know that the pressure difference is hydrostatic. Water does not flow from the bottom, where the pressure is greater, to the top because the pressure force is balanced by the body force of the fluid itself.

It is not pressure difference, Hubbert reasoned, but a potential difference which induces hydrodynamic circulation. Hubbert introduced the concept of *hydrodynamic potential*, which is defined by

$$\phi = p - \rho \, g \, H \; . \tag{9.11}$$

In a hydrostatic body of water, the pressure difference between the top and the bottom is the hydrostatic pressure, ρg H, but the potential difference is zero. Without a difference in the hydrostatic potential, no hydrodynamic movement can take place.

The hydrodynamic potential of laterally flowing groundwater at any point on land is represented by the height to which groundwater in a borehole will rise, i.e., the groundwater table (Fig. 9.4). The loci of the hydrodynamic potential from various sites constitute a potentiometric surface. Like electricity, water flows in the direction of maximum gradient on that surface.

Denoting the height of the potential surface at any point ground by ρg ΔH, groundwater flows because of the potential difference ρg $(\Delta H/\Delta L)$. Substitute the potential difference in place of pressure difference of Eq. (9.10), we have

$$q = k \, \frac{\rho g}{\eta} \left(\frac{\Delta H}{\Delta L}\right) \; . \tag{9.12}$$

This is the correct expression to relate the linear rate of groundwater flow to permeability (k) and hydraulic gradient, I propose calling this the *Darcy-Hubbert equation* in honor of this great scientist.

A comparison of the Darcy-Hubbert equation to the Darcy equation indicates

$$K = k \, \frac{\rho g}{\eta} \; . \tag{9.13}$$

Permeability (k) is not transmissibility (K). Transmissibility is a function of fluid viscosity. Permeability is a material constant; it should have the same value for a given piece of porous stone, regardless of whether air, water, or oil is to flow through the stone.

Permeability is, by definition, not a dimensionless number, but has the dimension of length squared. It is, as mentioned previously, a function of

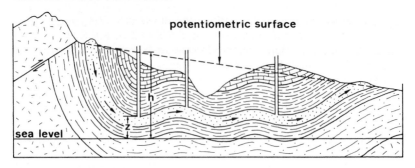

Fig. 9.4. Groundwater flow through an acquifer in humid regions. Regional flow of water through an aquifer sand from higher to lower outcrops shows a continuous drop of the hydrodynamic potential

the cross-sectional area of the conduit of fluid movement, or the average pore size of a rock.

The pore size of a rock is investigated by injecting mercury into its pore space: the smaller the pore space the more force is required to for the injection. The distribution of the pore size can be statistically evaluated, and average pore size calculated. Permeability is, however, not the pore size squared, but is related to that by the dimensionless number N. This number has been called a *tortuosity factor*. Interpreted within the framework of Poiseuille's law, fluid motion in porous medium can be thus compared to movement across a tortuous capillary tube of very small diameter D, with the transport rate related to the cross-sectional area, and the tortuosity of the path.

Permeability is not commonly expressed in square centimeters, but in darcies or millidarcies. A darcy is the permeability, in the c.g.s. system, of a medium which, according to the Darcy-Hubbert equation, permits the flow of a unit volume of a fluid, which has a unit viscosity, across a unit potential gradient. In terms of c.g.s. units,

$$1 \text{ darcy} = 0.987 \cdot 10^{-8} \text{ cm}^2 \approx 10^{-8} \text{ cm}^2 .$$

The reservoir sandstones in deeper zones of the Ventura Field have a permeability with values of several millidarcy, or 10^{-11} cm^2. That such sandstones can still make a good producing formation is explained by the fact that the hydrocarbons of Ventura have been "overpressured": The Ventura fluid pressure at 5000 m depth is more than twice the normal hydrostatic pressure at that depth. It was the great difference between the formation pressure and the borehole fluid pressure that permitted the outflow of oil at a reasonable rate.

Nevertheless, the permeability reduction with depth was a real problem, and the availability of data permitted a systematic study. Is the reduction of permeability caused by a change in sedimentary facies? Or is the trend an inevitable result of compaction at greater depth? Can the problem be evaluated quantitatively? Can the variation of permeability of reservoir beds in Ventura, or in any producing oil field, be predicted?

The first question is easy to answer because the permeability is related to the pore size, which, in turn, is related to grain size. Coarse sediments are good aquifers; they are permeable because the pore space between larger grains is, on the average, larger than that between finer grains. Better sorted sediments are also more permeable. Grains of very much smaller size could fall into the pore space between the larger grains and thus partially plug up the bigger pores. A well-sorted sediment, consisting of grains which are all about the same size, does not have such a problem; a relatively coarser grain cannot fall into a small pore. The permeability of an unconsolidated sediment is thus a function of its grain size and of the sorting. Where the former is commonly expressed by the median grain size, the latter is expressed by a dimensionless number σ_ϕ , which is a numerical expression of the phi class for the standard deviation. A value for $\sigma_\phi = 1$ signifies that two-thirds of all particles fall within one phi class.

The quantitative relation was investigated in 1947 by W. C. Krumbein and G. D. Monk, and they found

$$k = C_0 \, D_m^2 \, e^{-c \, \sigma_\phi} \, , \tag{9.14}$$

where D_m is the median diameter of the sediment, C_0 and c are two experimental constants which have been determined to be 760 (darcy/mm^2) and 1.31, respectively.

Using the relation of Krumbein and Monk, we can compute the permeability of any sedimentary rock by substituting the values of d_m and σ_ϕ , which are determined by sieve analysis. I did that and found that the reduction of permeability at Ventura cannot be related to changes in sedimentary facies. The turbidites of Ventura from each zone have the same wide range of size and of sorting (Fig. 9.5), but not the same permeability range (Fig. 9.6). The difference in the permeability values of the shallower and the deeper rocks must thus be related to the process of *lithification*

Lithification, meaning changing into rock, is the process which makes a solid rock out of loose sediment. Such processes, which take place after the deposition of sediments are also called *diagenesis*. In detrital sediments, the two major diagnetic processes are cementation and compaction.

Cementation by calcite can plug up almost all the space and produce a non-porous and non-permeable sandstone. This type of sandstone exists in Ventura, but it constitutes only a minor bulk of the reservoir beds. The

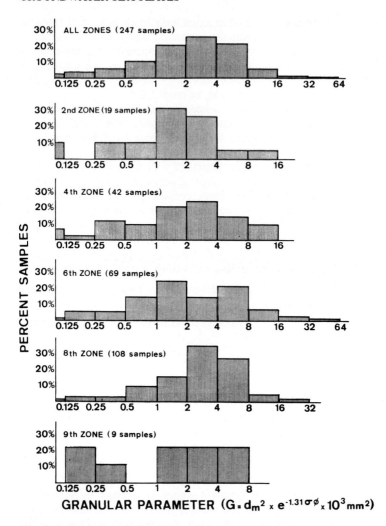

Fig. 9.5. Variation of lithology of Ventura sediments (after Hsü 1977). The permeability of an unconsolidated sediment can be calculated on the basis of an empirical formula 760 x granular permeability. This figure shows that the ranges of the values of the granular parameter of the sediments from various zones are about the same. This permits the conclusion that the permeability decrease with depth at Ventura has been caused by increased compaction

Ventura Field, California, II

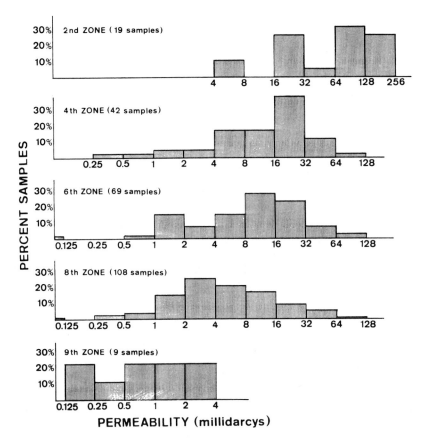

Fig. 9.6. Range of permeability, Ventura field (after Hsü 1977). The permeability of sandstones is on the whole much less in deeper formations of Ventura. This figure shows the percentage distribution of specimens of various permeability. In the shallower 2nd zone, most samples have between 64 and 128 millidarcy, but the mode of the 8th zone samples is between 2 and 4 millidarcy

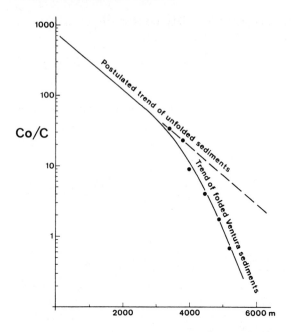

Fig. 9.7. Compaction trend at Ventura (after Hsü 1977). The ratio of the measured permeability of a lithified sediment and the calculated permeability when it is still unconsolidated, C/C_0 gives a measure of the degree of compaction of the Ventura sandstones. An empirical trend of permeability reduction as a function of overburden at the time of deformation has been obtained, and this trend serves to predict the permeability reduction for sandstones of various types at greater depth

great majority of the sandstones have been lithified because of compaction (Fig. 9.7). Under the pressure of a heavy overburden and deformed by shearing stress, the loose grains have been rearranged so that the pores have become smaller thus reducing the permeability. The measured permeability k of a sediment is compared to the calculated value k_0 of the sediment in an unconsolidated state, and a *compaction index* is obtained

$$C / C_0 = k / k_0 = \text{compaction index} . \tag{9.15}$$

The smaller the ratio, the more compacted is the sandstone. I found that the average value of the ratio is 0.045 for samples in the uppermost Pliocene reservoirs and 0.0009 for those in the deepest Pliocene reservoirs of Ventura. In other words, the permeability has been reduced to 4.5% of the original in the former, and to less than 0.1% of the original in the latter.

With the concept of the compaction index, the degree of compaction of any sedimentary rocks in any region can be numerically represented, and the permeability reduction can be statistically evaluated. An estimate of the permeability that could be expected from each type of rock can be predicted. This methodology has found its way into practice.

The results of my Ventura studies demonstrate the importance of compaction in lithification. Compaction is a particularly important process because the Ventura sediments are poorly sorted and muddy. The poor sorting permits the smaller grains to plug the holes between the bigger grains, and the muddy minerals are deformable to mold themselves between more rigid grains. Some sandstones, such as the St. Peter Sandstone of Cambrian age in the United States, consist of quartz spheres of about the same size. Rearrangement by compaction does little to reduce the porosity and permeability of such sediments. In such cases the lithification of sediments often results from chemical reactions.

Chemical changes involve the addition to (cementation), the removal from (dissolution), or the exchange with (recrystallization or replacement) the material in the original sediments. Cementation, as we have mentioned previously, plugs up the pores. Dissolution removes the irregularities or protrusions of misfitting grains to make a compact rock. Recrystallization or replacement causes a new arrangement of newly formed crystals. If they all fit together like a mosaic of stones, they plug up the holes. If they form an aggregate of euhedral crystals, such as *sucrosic dolomite*, which is composed of rhombs, they could be very porous and permeable, therefore dolomite is a well known oil reservoir in carbonate terranes.

The addition, removal, or exchange of chemical components in a system requires material transport. How do chemical components get displaced. The two processes are chemical diffusion and hydrodynamic transfer.

Diffusion of dissolved ions in the pore solution of a sediment is the only effective mode of material transport in a hydrostatic situation, i.e., the quantity of a component moving in the direction of a concentration gradient, from the more to the less concentrated.

An understanding of the role of diffusion has led me to tackle the question on the origin of chert. Cherts form nodular bodies or lenticular layers in another host rock, which is commonly a limestone or a radiolarite. Yet such chert nodules or lenses have never been found in Recent sediments.

In 1968, I joined a deep-sea drilling expedition to the South Atlantic. On the west side of the Mid-Atlantic Ridge, we drilled into soft radiolarian oozes of Eocene age. The skeletal debris of radiolaria are composed of amorphous silica. The drilling rate was very fast, up to a 100 m/h, through the ooze. Then a hard layer was hit. When the core finally came up after hours of grinding, we found that we had drilled through a thin chert layer, which was embedded in the soft ooze.

Why should most of the silica be preserved as fossil debris in the oozes, yet some have been recrystallized to form chert?

The water in bottom sediments does not move; deep-sea sediments are commenly hydrostatic. The water chemistry suggests, however, upward movement of various saturated components. The ions in the water of the pore space move by diffusion. In areas underlain by radiolarian or diatom oozes, the pore water equilibrating with the amorphous silica in the sediment has a silica concentration many times that of the seawater, which is undersaturated in silica. Silica moves through the pores of the sediments by ionic diffusion, because of the presence of this concentration gradient. The movement rate is

$$q = c \left(\frac{d\ddot{C}}{dx} \right) .$$

where q is the rate of material transport per cross-sectional area, c the diffusion coefficient of silica, and $\left(d\ddot{C} / dx \right)$ the concentration gradient.

These results caused me to consider. I knew that the various silica compounds have different "apparent solubilities". Dissolution of amorphous silica at room temperature could yield a solution with more than 100 ppm SiO_2, yet a saturated solution of dissolved quartz has a concentration of less than 10 ppm. If a sediment contains both forms of silica, a concentration gradient is set up between the two. The pore water surrounding amorphous silica grains, such as radiolarian skeletons, has more than ten times the SiO_2 concentration as that surrounding a quartz grain. Dissolved silica should then move toward the quartz by ionic diffusion. This causes a supersaturation with respect to quartz. Silica is deposited on the surface of the quartz grain, which now becomes a center of nucleation.

Diffusion is a very slow process. The diffusion coefficient for SiO_2 is only 1.5×10^{-6} cm^2/s under sea bottom conditions. With an optimum gradient of some 40 ppm/m, the flux by diffusion is only 2×10^{-5} g/cm^2 per year. No wonder we have seen practically no chertification of Recent sediments; the duration of the last 10000 years is much too short for extensive material transport by diffusion. Cherts are present in Miocene ocean sediments of 5 to 6 million years of age, and are common in older Oligocene and Eocene sediments.

A more effective means of material transport is the movement of subsurface waters. Such movement has been termed *hydrodynamic* , and knowledge of hydrodynamic circulation is indispensable for an understanding of diagenetic processes involving extensive material transport, processes such as cementation, dissolution, and replacement.

The key to the understanding of diagenesis is to realize that groundwater flow is a physical process. Groundwater is a chemical solution which loses its chemical equilibrium when it is transferred to a new environment. Diagenesis is an attempt to re-establish this equilibrium. From this viewpoint, chemical diagenesis is an inevitable consequence of hydrodynamic circulation. We shall come back to this question when we discuss the origin of dolomites.

Suggested Reading

M. King Hubbert's *Theory of ground-water motion* (J Geol 48:785-944, 1940) should be read by all geology students. In this paper, Hubbert discussed the fundamental experiment by Henry Darcy, described in *Les fontaines publiques de la ville de Dijon* (Paris: Victor Dalmont, pp 590-594, 1856), and pointed out the sloppy thinking by hydrologists of the modern era who neglected the gravity term in the equation of groundwater motion.

My ideas on material transfer in diagenesis through ionic diffusion, as exemplified by chertification, have been explained in my 1976 article *Paleoceanography of the Mesozoic Alpine Tethys* (Geol Soc Am Spec Pap 170:27-36). My work on Ventura was first written in 1955 as a report of the Shell Development Company and subsequently published in 1977 and 1979 as articles in the bulletin of the American Association of Petroleum Geologists (61:137-168; 169-191;64:1034-1051).

10   Components Equilibrate

Mass-Action Law - Gibbs Criteria of Chemical Equilibrium
First and Second Law of Thermodynamics
Lewis Concept of Chemical Activity - Carbonate Equilibria
Metastable Phases - Calcite Dissolution

Chemical sediments are precipitates from natural waters. Solids precipitate from solution when the solution is saturated or supersaturated, and a solution is saturated when the concentration of the solutes exceeds the saturation value, as defined by the mass-action law in middle-school chemistry:

$$(C^c)\,(D^d) \,/\, (A^a)\,(B^b) = K \; , \tag{10.1}$$

where K is an equilibrium constant, constant at a given temperature and pressure, and (A), (B), (C), (D) are concentrations of chemical components in an ideal solution. The values of a, b, c, d are defined by the chemical reaction

$$a\,A + b\,B = c\,C + d\,D \; . \tag{10.1a}$$

The mass-action law was formulated by two Norwegian scientists, C. M. Guldberg and Peter Waage in 1864, on the basis of experiments with very dilute solutions. We are told that the mass-action law is only valid for ideal solutions of nearly zero concentration. Natural waters, such as seawater or saline groundwater, are not ideal solutions. For those, the symbols (A), (B), (C), and (D) in Eq. (10.1) refer to the *chemical activities, or activities,* rather than concentrations of the species in solution.

The meaning of activity is not self-evident to undergraduates, and many students in my class have asked for explanations. My usual reply, which may not please a theoretical purist, is to equate activities to "effective concentrations". We know that the deviation from the mass-action law becomes more substantial as the solution becomes more and more concentrated. The concept activity is an invention to rescue the mass-action law. To retain the validity of Eq. (10.1), the values (A), (B), (C), (D) cannot be the concentrations as such; they have to be values of the experimentally determined concentrations multiplied by a correction factor, which is called *activity coefficient.* Experiments have shown that the values of the activity coefficient of chemical species are related to the total concentration of the solution, as expressed in *ionic strength*, which is the summation of ionic concentrations, and to other factors such as the diameters of the ions in solution and of their valence. The values of activity coefficients are commonly less than one, signifying the effective concentration of a solute in non-ideal solutions is less than the measured concentration.

Geochemists have used an empirical relation, the Debye-Huckle formula deduced by the two chemists of those names, to calculate the values of activity coefficients of chemical species in natural waters. We can, therefore, make theoretical predictions as to whether natural waters are saturated, undersaturated, or supersaturated with the solid phases, to which they are in contact. Saturation is an equilibrium state, but a slight deviation from that state would cause chemical changes: Undersaturation causes dissolution, supersaturation causes precipitation.

For those readers of this book who are prepared to accept the mass-action law as an axiom, we could proceed from middle-school chemistry to interpret geochemical phenomena. They could skip the next section on chemical thermodynamics. I have written a derivation of the mass-action law for students, such as those in our university, who have to take a required course on physical chemistry. With their knowledge of basic principles of thermodynamics, they could appreciate that chemical processes are, in a broad sense, physical processes, and they could understand that the mass-action law is an experimental verification of Gibbs' theoretical criteria of chemical equilibrium.

Relations in natural sciences are expressed by equations, and equations are written to express equilibria. The equations in the first nine chapters of this book are mainly expressions of equilibria of forces, which are manifestations of the First and Second Newtonian Laws of Mechanics. Chemical equilibria are thermodynamic, meaning that the role of heat in the flux of energy and matter is taken into consideration.

J. Willard Gibbs, a great American chemist of the last century, has stated the general conditions of chemical equilibrium in terms of energy fluxes in a system consisting of more than one homogeneous part:

$$dE' + dE'' + dE''' + \ .... \ = 0 \ . \tag{10.2}$$

The terms dE', dE", and dE''' are the variations of the energy within three homogeneous parts of the system; the letters referring to the different parts are designated by different accents. The variation of any homogeneous part of variable chemical composition of a given mass is defined by

$$d E = T \ dS - p \ dV + \mu_1 dm_1 + \mu_2 dm_2 + ... + \mu_n dm_n \ , \tag{10.3}$$

where E denotes the total energy of the homogeneous part, T its absolute temperature, S its entropy, p its pressure, V its volume; $m_1, m_2, ... m_n$ are quantities of the various substances, and $\mu_1, \mu_2, ... \mu_n$, denote the chemical potentials of the various substances, or the differential coefficients of energy content, dE taken with respect to $m_1, m_2, ... m_n$. The variables for the various phases are denoted by the primes, e.g. T', T", T"' are the symbols for the temperatures of the first, second, and third phases. Substitute Eq. (10.3) into Eq. (10.2), we have

$$\begin{aligned}
&\text{T'dS' - p'dV'} + \mu_1{}' \, dm_1{}' + \mu_2{}' dm_2{}' + ... + \mu_n{}' dm_n{}' \\
&+ \text{T"dS" - p"dV"} + \mu_1{}" dm_1{}" + \mu_2 dm_2{}" + ... + \mu_n{}" dm_n{}" \\
&+ \text{T"'dS"' - p"'dV"'} + \mu_1{}"' dm_1{}"' + \mu_2{}"' dm_2{}"' + ... + \mu_n{}"' dm_n{}"' = 0 \; .
\end{aligned} \tag{10.4}$$

A solution of Eq. (10.4) is

$$\begin{aligned}
&T' = T" = T"' \\
&p' = p" = p"' \\
&\mu_1{}' = \mu_1{}" = \mu_1{}"' \\
&\mu_2{}' = \mu_2{}" = \mu_2{}"' \\
&\mu_n{}' = \mu_n{}" = \mu_n{}"' \; .
\end{aligned} \tag{10.5}$$

Equations (10.5) are the necessary and sufficient conditions as defined by Gibbs in 1878.

Gibbs' equations seem quite complicated. I am not foolhardy enough to teach you chemical thermodynamics in one lecture on introductory sedimentology. On the other hand, I would like to remain true to my promise not to throw equations at you, without explaining the physical basis of their derivation. Gibbs' definition of equilibrium can be considered a restatement of the First and Second laws of thermodynamics. What are the physical bases for the formulation of these two fundamental concepts?

The First Law of Thermodynamics relates energy, work, and heat; all these are physical entities.

In our discussion of rockfalls, we have stated that the potential energy of a falling rock mass is partly converted into the kinetic energy of the flowing debris and partly dissipated by friction, or

Potential energy = kinetic energy + friction .

The kinetic energy of the flowing debris is

$$\text{K.E.} = \frac{1}{2} M u^2 \; . \tag{10.6}$$

The velocity of the flow, starting from stationary rest, is

$$u = s / t .$$
(10.7)

The distance of travel, L, by the flow is

$$s = \frac{1}{2} a t^2 .$$
(10.8)

Substitute Eqs. (10.7) and (10.8) into Eq. (10.6), we have

$$K.E. = \frac{1}{2} Mu^2 = M \, a \, s = F \, s$$

or

Kinetic energy $=$ force $\cdot$ distance $=$ work .
(10.9)

What happened to the energy dissipated by friction? It is turned into heat and is transferred away from the flowing debris. The conservation of energy as applied to the physics of rockfall states that the decrease of potential energy of a rockfall, -dE, is partly converted into work done by the moving debris, dW, and partly into heat, -dQ, which is dissipated and lost to the system of moving debris.

or

$$- dE = dW - dQ .$$
(10.10)

Rearrange the terms, we have

$$dE = dQ - dW .$$
(10.11)

This is, of course, the First Law of Thermodynamics, which evaluates not only the work done by or done to the system, but also the heat absorbed or released in the system of changing energy content.

In chemical reactions, not only heat or work but chemical energy must be considered. Since the chemical potential μ has been defined as the differential coefficient of energy content, dE, taken with respect to mass, dm, the change in chemical energy of the total mass of a system is thus the integration of chemical potential, or μ dm. When the chemical change is considered, the law of conservation of energy states thus

$$d E = d Q - d W + \mu \, d m .$$
(10.12)

In systems involving chemical changes, the mechanical work done is

$$W = F . L = p A \; dL = p \, dV ,$$ (10.13)

where p is pressure and A the cross-sectional area upon which the pressure acts.

The change of heat is related to entropy, which has been defined as the heat absorbed by a system in a reversible process, dQ, divided by the temperature:

$$dS = \frac{dQ}{T} .$$ (10.14)

This definition is, of course, the Second Law of Thermodynamics; you should have learned that in your chemistry course if not already in middle school. We have, therefore, for reversible processes, i.e., at equilibrium:

$$dQ = T \, dS .$$ (10.15)

Substitute Eqs. (10.13) and (10.15) into Eq. (10.12), we have

$$dE = T \, dS - p \, dV + \mu \, dm .$$ (10.16)

The total change of chemical energy involving components with masses $m_1, m_2, ...$ is the sum of the changes of all the components in all the phases

$$\mu \, dm = \mu_1{}' \, dm_1{}' + \mu_2{}' \, dm_2{}' + ...+ \mu_2{}''dm_1{}''+ \mu_2{}'' \, dm_2{}'' + ...$$ (10.17)

Substitute Eq. (10.17) into Eq. (10.16), we have the equation for chemical equilibrium as given by Gibbs (Eq. 10.3).

Considering the chemical equilibrium, as expressed by the relation (10.1a), involving one solution phase and four components at temperature T and pressure p, Eq. (10.3) can be simplified to read

$$dE = TdS - pdV + \mu_1 dm_1 + \mu_2 dm_2 - \mu_3 dm_3 - \mu_4 dm_4 ....= 0 .$$ (10.18)

In this case, it can be stated that at the temperature T, pressure p, the sum of chemical energy fluxes is null at equilibrium, or

$$\mu_1 \, dm_1 + \mu_2 \, dm_2 = \mu_3 \, dm_3 + \mu_4 \, dm_4 .$$ (10.19)

Equation (10.1) is a statement of chemical equilibrium in terms of ionic concentrations in ideal solutions, and Eq. (10.19) is a statement in terms of chemical potentials (also known as partial molar free energies). These two equations are made identical through the introduction of the concept of activity by the noted American chemist G. N. Lewis, who defined in 1923 chemical activity $\underline{a}$ in terms of chemical potential by the equation:

$$\mu = \mu_0 + RT \ln \underline{a} \qquad (10.20)$$

$$\lim (\underline{a}/c) = 1, \text{ for } c \to 0 \quad ,$$

where μ_0 is the chemical potential of a dissolved component at a standard state and is a constant at a given temperature and pressure, and c is the ionic concentration. You can see from this equation that chemical potential is energy (expressed in calories or joules) per unit mass (expressed in grams or moles), and activity is dimensionless (commonly expressed in weight molar concentration).

For a system involving one homogeneous phase of four dissolved components in solution, as expressed by the relation (10.1a), we have, from Eq. (10.20):

$$\mu_A = k + RT \ln (A) \qquad (10.21)$$

$$\mu_B = k + RT \ln (B)$$

$$\mu_C = k + RT \ln (C)$$

$$\mu_D = k + RT \ln (D) .$$

The changes of mass in the reaction indicated by Eq. (10.1a) are expressed by a, b, c, d. Substitute the expressions given by Eq. (10.21) to Eq. (10.19), we have, for a given temperature and pressure,

$$- a \ln (A) - b \ln (B) + c \ln(C) + d \ln (D) = k . \qquad (10.22)$$

Rearrange the terms, we have

$$(C)^c (D)^d / (A)^a (B)^b = k .$$

This is the mass-action law.

I have made an excursion to derive theoretically the mass-action law to demonstrate that chemical processes are physical processes. The science of chemistry differs, however, from the study of mechanics in that not only work is considered, but also fluxes of heat and changes of chemical energy, which are related to entropy and chemical potentials, two basic concepts in chemical thermodynamics.

Sedimentary processes involving chemical changes are precipitation, dissolution, and replacement. Precipitation occurs in saturated, and dissolution in undersaturated solution. Replacement of one mineral (a solid phase) by another commonly involves equilibrium with a solution (the third phase) which is undersaturated with respect to the former and supersaturated with respect to the latter.

In the case of calcite precipitation, for example, the process takes place, according to mass-action law, when the condition is

$$(Ca^{2+})(CO_3^{2-}) = K_{CaCO_3} . \qquad (10.23)$$

The left side of the equation is the activity product of calcium and carbonate ions in solution, or the IAP. The right side of the equation, K_{CaCO_3} or simply K_c, is the solubility of calcite, which is a constant at a given temperature and pressure. If the ionic activity product of a solution exceeds the solubility, the solution is supersaturated, calcite should precipitate. If IAP is less, the solution is undersaturated, calcite should be dissolved.

The dissoved calcium of natural waters is derived from weathering of rocks, the concentration of calcium ions in solution does not vary very much under ordinary conditions. This is not the case of the dissolved carbonate ions, the concentration of which could vary greatly as a consequence of varying temperature or hydrogen ion concentration. Calcite precipitation in natural waters is thus normally related to changing carbonate-ion concentration, or more precisely, carbonate-ion activity.

Dissolved carbonate ions are only one of the species when carbon dioxide is dissolved in a solution (Fig. 10.1); the other species are HCO_3^-, dissolved CO_2, or $CO_2(aq)$. The total dissolved carbonates are equilibrated with the carbon dioxide in air, or $CO_2(gas)$, and the relative activities of the various species are governed by the considerations of the following reactions:

$$CO_2(gas) = CO_2(aq) \qquad (10.24a)$$

$$CO_2 (aq) + H_2O = H_2CO_3 \qquad (10.25a)$$

$$H_2CO_3 = H^+ + HCO_3^- \qquad (10.26a)$$

$$HCO_3^- = H^+ + CO_3^{2-} . \qquad (10.27a)$$

157

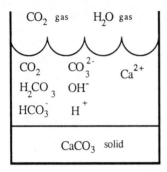

Fig. 10.1. $CaCO_3$-H_2O system. The three phases of the two-component system are the gas, solution, and the solid phases. The components in the gas phase are CO_2 and H_2O, and the only one in the solid is $CaCO_3$. The solution phase consists, however, of numerous dissolved species, and their relative concentration is governed by Henry's law and by the equilibria governing the dissociation of carbonate ions

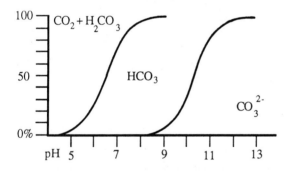

Fig. 10.2. Variation of the carbonate-ion concentration with pH. Dissolved carbon dioxide is the dominant species in acidic, bicarbonate ions in nearly neutral and carbonate ions in alkaline solutions

The equilibrium constants are:

$$(CO_2)/ p_{CO2} = K_{CO2} \qquad\qquad (10.24)$$

$$(H_2CO_3)/(CO_2)(H_2O) = K_{H2CO3} \qquad\qquad (10.25)$$

$$(H^+)(HCO_3^-)/ (H_2CO_3) = K_1 \qquad\qquad (10.26)$$

$$(H^+)(CO_3^{2-}) /(HCO_3^-) = K_2 \; . \qquad\qquad (10.27)$$

Equation (10.24) is an expression of Henry's law which we learned in middle-school chemistry: the concentration of dissolved gas in solution is directly proprotional to the partial pressure of the gas in air. The equilibrium constant is temperature dependent and is found experimentally to have a smaller value at higher temperature. An impulsive person might rush to the conclusion that a temperature increase would cause a decrease of dissolved carbon dioxide, and thus a decrease of carbonate ions, leading to the condition of calcite undersaturation. The first part of the statement is correct, according to Eq. (10.24), but the second part of the statement is paradoxically incorrect. A decrease of total dissolved CO_2 does not decrease, but increases the carbonate-ion concentration. Why?

A consideration of Eqs. (10.25) (10.26) and (10.27) shows the role of varying pH in calcite precipitation. Carbon dioxide dissolved in water becomes carbonic acid. The dissociations of carbonic acid are described by the reactions (10.26a) and (10.27a), and the equilibrium constants are called first and second dissociation constants of carbonic acids. Rearranging the terms of Eqs. (10.25), (10.26), and (10.27), we have

$$(CO_3^{2-}) = k (CO_2)(H_2O)/ (H^+)^2 \; . \qquad\qquad (10.28)$$

The activity of water of a dilute solution is unity, or approximately so, because a kilogram of solution is, for all practical purpose, made up of a kilogram of water. Assuming a value of unity for (H_2O) of natural waters, Eq. (10.28) is thus simplified:

$$(CO_3^{2-}) = k (CO_2)/(H^+)^2 \; . \qquad\qquad (10.29)$$

In other words, the carbonate-ion activity of a solution varies linearly with the dissolved carbon-dioxide activity, but is inversely proportional to the square of the hydrogen-ion activity. The removal of dissolved carbon dioxide from a solution renders it less acid. With the hydrogen-ion activity reduced, or pH increased, the solution becomes more alkaline and the bicarbonate and carbonate-ion concentrations are increased. The distribution of the various carbonate species in a solution as a function of pH is calculated on the basis of Eq. (10.29) and is diagrammatically illustrated by Fig. 10.2. The total dissolved carbon carbonates consists mainly of dissolved CO_2 in very acid solutions with a pH of 5 or less. The carbonate-ion CO_3^{2-} becomes the dominant species in alkaline solutions with a pH greater than 10. In most natural waters the pH values are not so extreme and the bicarbonate ions constitute the most common species.

The precipitation of calcium carbonate can thus be caused by an increase in carbonate-ion activity as a consequence of removing dissolved CO_2 from natural waters. Such removal could be related to a temperature change: cold water ascending from the depth to the surface may become warmer and thus loses dissolved CO_2. Coating of agitated grains forming *oolite* in tidal channels of the Bahamas is such an inorganic precipitation of calcium carbonate from supersaturated seawater.

Dissolved CO_2 in natural waters can also be removed because of its utilization by photosynthetic organisms. Geochemical studies revealed that the seawater is highly supersaturated in areas where the shallow bottom is densely populated by green algae. Photosynthesis led to depletion of total dissolved carbonates (Fig. 10.3). The pH of surface seawater ranges from 8.1 to 8.3 and is more alkaline than the intermediate and bottom waters, which have a pH of 7.8 or 7.9. The carbonate-ion concentration is increased, when carbon dioxide is utilized by photosynthetic organisms. Calcium carbonate, in this case, is inorganically precipitated as an indirect result of biotic activities.

Calcium carbonate forms the internal or external skeletons of many marine and terrestrial organisms, and such skeletons are in fact a major source of debris for carbonate sediments. The biogenic carbonate, however, is not necessarily formed under equilibrium conditions for inorganic precipitation, and the departure from equilibrium is called the vital effect of living organisms.

Two minerals of calcium carbonate occur in nature: calcite and aragonite. They have different crystal structures, and could thus take on different traces of impurities. The calcite structure is rhombohedral, sufficiently similar to that of magnesite so that $MgCO_3$ can be present as a solid solution in calcite. A low magnesium calcite contains only a few mol% of magnesium carbonate, but a high magnesium calcite may contain up to 10%. The aragonite structure is orthogonal like strontianite and thus commonly contains Sr as a trace element.

If you drop a calcite crystal in one beaker full of distilled water and an aragonite in another, the concentrations of the dissolved ions are not the

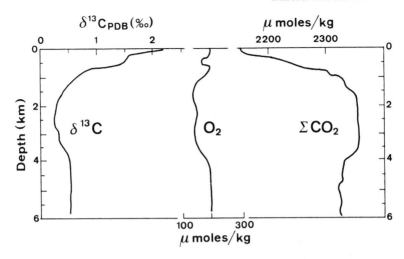

Fig. 10.3. Vertical gradient of seawater chemistry. The surface seawater is depleted in total dissolved carbon dioxide because of photosynthetic activities of oceanic planktons. Oxidation of decayed organisms at intermediate depth causes depletion of oxygen below the surface. The oxygen content is minimum at an intermediate depth, where the pH of seawater also has a minimum value of 7.7. The curve on the *left* shows the vertical gradient of the carbon isotopes, which is also a record of fractionation by the photosynthesis of oceanic planktons, a process to be discussed in the next chapter

same, even if the two beakers have been kept sealed under the same temperature and pressure. More calcium carbonate is dissolved in the aragonite beaker.

If sufficient time for dissolution is allowed, the final concentration of the solution in the calcite beaker is called the solubility of calcite. At such a stage, we can say that a chemical equilibrium between the solution and the solid phases has been reached, or the condition as defined by Eq. (10.23) is fulfilled. The equilibrium activity product of the calcium and carbonate ions in solution (IAP) is the solubility of calcite at the temperature and pressure where the sealed beaker full of solution is kept.

The final concentration, or the IAP, of the components in the aragonite beaker has often been called the "solubility" of aragonite. A purist would frown upon such a practice, because aragonite is not a stable phase under atmospheric conditions, it cannot be in chemical equilibrium with a solution phase at room temperatures and pressures. Theoretically, as the aragonite dissolves and releases calcium and carbonate ions, their activity product should exceed the equilibrium IAP or K_c, the concentration of the

161

ions in solution should be reduced by precipitation of calcite. As aragonite should continuously be dissolved to supply more calcium and carbonate ions to the solution, we should not have any aragonite solid left when equilibrium is established, but calcite and its saturated solution. In actuality, calcite may not be precipitated in the aragonite beaker. Aragonite would stop dissolving when a maximum concentration is reached. We can say that a metastable equilibrium is reached, and that the maximum concentration or IAP could be referred to as the metastable solubility or simply "solubility" of aragonite. The "solubility" of a metastable phase is always greater than the equilibrium solubility of a stable phase.

Aragonite is a common mineral in Recent sediments. It is the main constituent of the skeletons of numerous organisms. The molluscan shells are, for example, commonly aragonite. Their occurrence as biogenic debris is explained by the fact that vital organisms do not always carry out their biochemical reactions under chemical equilibrium conditions. Yet aragonite is also a common chemical precipitate. Aragonite coating of detrital nuclei forms particles called *ooids*, and an oolite is a sediment made exclusively of such ooids. Furthermore, the cementation of marine sediments consists, as a rule, of aragonite. The common occurrence of aragonite, instead of the stable phase calcite, in Recent marine sediments indicates that chemical equilibrium can only define the theoretical tendency. What actually takes place is, in many instances, governed by the rules of chemical kinetics.

Supersaturation is possible because the establishment of equilibrium is not instantaneous. Natural waters can become quickly supersaturated; the seawater near the surface is, as a rule, supersaturated with calcite, but its precipitation is commonly hindered: The presence of various salt ions in seawater apparently interferes with the ordering of calcium and carbonate ions so that no calcite lattice can be formed. When seawater supersaturation reaches above the aragonite "solubility", aragonite precipitates; the interference of other ions does not effectively prevent the crystallization of aragonite. Consequently, the common chemical precipitate from seawater is this metastable phase. Eventually aragonite in marine sediments is changed to calcite by diagenesis, and ancient limestones consist almost exclusively of calcite.

Supersaturation of lake waters is also a common phenomenon. In our study of Lake Zurich chalk, Kerry Kelts and I found that the IAP of the lake water can be ten times more than the solubility. Instantaneous equilibrium is hindered because calcium carbonate is not being precipitated fast enough to reduce the carbonate-ion concentration in lake water. Nevertheless, the chemical kinetics of freshwater precipitation is such that the stable phase calcite is precipitated.

The surface seawater, which is equilibrated with the atmosphere, is depleted in carbon dioxide due to the photosynthesis of oceanic planktons.

The deeper water contains more carbon dioxide because of the decay of dead organisms. Such decay consumes oxygen, and the CO_2 concentration is maximum in the oxygen-minimum zone (Fig. 10.3). At greater depth, the bottom water is also enriched in carbon dioxide, which has a partial pressure three times as large as its partial pressure in the atmosphere.

Calcite is more soluble in CO_2-rich waters. Supersaturation changes to undersaturation at an oceanic depth of several hundred meters. Yet calcitic sediments are found at seafloor down to a few thousand meters, where the seawater is undersaturated, because dissolution is not instantaneous.

The kinetics of chemical reaction is dependent upon reaction temperature and degree of supersaturation. Dissolution of calcite is faster in deeper waters because of its greater departure from equilibrium, even though colder temperatures usually retard reaction rate. Mel Peterson suspended calcite spheres in ocean water and found that the dissolution rate increases rapidly at a depth of a few thousand meters, which varies from place to place. The surface below which accelerated dissolution takes place is called by Wolfgang Berger *lysocline*. This word is a parallel to the term *thermocline*, which denotes the imaginary inclined surface in a natural water body, at which the temperature changes rapidly.

At some depth below the lysocline, the reaction rate is so fast that the supply of calcitic sediment is completely dissolved. This depth also varies from place to place, and is known as the *calcite-compensation depth,* or *CCD*. Compensation means that the dissolution of calcite at that depth is exactly compensated by the supply of calcite settling out of the overlying water column. Berger compared the CCD to the snowline. Imagine an ocean without water, you could visualize deep basins underlain by red clays. Higher above on swells or oceanic plateaus are the calcite oozes, shiny white in color, covering the highs like snow in the mountains. The CCD ranges from a few thousand to more than 5000 m oceanic depth. From the investigations of the deep-sea sedimentary record during the last two decades, we now know that the CCD has varied with time. This variation is indicated by the changing lithology of an oceanic sequence, which may vary from pure ooze to marly ooze to red clay even in regions where the ocean bottom did not move up or down very much.

Aragonite is more soluble than calcite. The aragonite-compensation depth (ACD) lies higher in than the CCD in the ocean. There is an interval between the ACD and CCD in which all aragonite is dissolved, but some calcitic sediment remains. The famous *aptychus* limestone of the Alps was obviously deposited in such a zone. Aptychus is the calcitic cover of the aragonitic shell of an ammonite. Fossil aptychus debris form lime sediments at a depth below the ACD, where the aragonitic shell is completely dissolved, but above the CCD.

Calcium carbonate is not the only solid which builds up more than one naturally occurring mineral of that composition, silica also occurs in several forms in nature. Biogenic silica is commonly amorphous. Some chert can be identified as a fine-grained mixture of cristobalite and tridymite, which are stable phases at higher temperatures, such chert is called *opal C-T*. Quartz is the only stable phase of SiO_2 at room temperature and atmospheric pressure. Like all amorphous or metastable phases, the "solubility" of biogenic silica is less than the solubility in quartz. In a sediment containing both amorphous silica and quartz, the concentration of dissolved SiO_2 surrounding biogenic debris can be about ten times as high as that surrounding a quartz grain. A concentration gradient can thus be set up, so that solute is transported across the gradient. Finally, all the amorphous silica is dissolved and quartz is accreted on a nucleus to form a nodule, or layer. This origin of chert has been discussed in the Chapter 9.

We have so far discussed diagenetic problems involving no exchange of materials across the system boundaries, such as the closed systems. Chemical alteration of sediments, or chemical diagenesis, involves, in numerous instances, materials added to and/or substracted from a system. Diagenesis in such an open system requires an agent for material transport. Diffusion is not an effective mechanism for material transport, but hydrodynamic circulations of subsurface waters are. A basic understanding of the principles governing such circulations is, therefore, a prerequisite for an understanding of diagenesis. Unfortunately, theory of groundwater flow has been considered by many a specialty for hydrologists and is not taught everywhere in courses on sedimentology. We have discussed the physical principles of groundwater movement in the last chapter and shall discuss an application of those principles to problems of chemical diagenesis in the next.

Suggested Reading

The first sections of this chapter are a partial summary of my 1967 paper on *Chemistry of dolomite formation* (in *Carbonate rocks*, G. V. Chilingar et al. (eds) Development in Sedimentology, 9B. Amsterdam: Elsevier, pp 169-191). For graduate students in geochemistry who wish to delve deeper into the fundamentals, they would want to read Gibbs' original monograph *On the equilibrium of heterogeneous substances*; the Dover Publications (New York) have reprinted *The scientific papers of J. Willard Gibbs* in 1961. Students seeking to understand better the concept of activity should consult L. H. Adams' article on *Activity and relatd thermodynamic quantities; their definition, and variation with temperature and pressure* (Chem Rev 19:1-26, 1936).

Although numerous and excellent books have been written on the chemistry of seawater in recent years, I still find the treatment in *The oceans*, by H. U. Sverdrup, M. W. Johnson and R. H. Fleming (New York: Prentice Hall, 1942), the most enlightening for beginning students. The principles of marine geochemistry have been applied to many sedimentological studies, too numerous to be cited. I might recommend the book which I edited with Hugh Jenkyns on *Pelagic sediments* (Int Assoc Sediment Spec Publ No 1, 1974); an excellent article on *Plate stratigraphy and the fluctuating carbonate line* by W. H. Berger and E. L. Winterer is published in that volume.

11   Evaporation Pumps

Dolomite - Dolomite solubility - Material Transfer in Open Systems
Seepage Refluxing - Evaporative Pumping - Flood Recharge

Dolomite is a mineral consisting of calcium and magnesium carbonates of about 50 mol% each. Dolomite is also the name of a rock made up of the mineral dolomite. The origin of dolomite has been a puzzle for many years, because the carbonate deposits of Recent environments are commonly lime sediments; yet ancient carbonate sequences are, in places, predominantly dolomite. Recent dolomite occurrences are rare; Recent dolomite was first positively identified in Australia during the late 1950s. Since then, several other occurrences have been discovered: in the Persian Gulf, on Caribbean islands, in Equatorial Pacific lagoons, on intermontane playas, etc.; all these places share one common characteristic: their climate is arid.

Some dolomite is very fine grained and occurs in thin laminae. Such cryptocrystalline dolomite has been called "primary dolomite", although it is still debated among specialists whether dolomite can ever be precipitated from natural waters. Most dolomites are secondary, formed by replacement. Sedimentary structures in Recent carbonate sediments are commonly preserved in dolomites so that the postulate of their secondary origin is commonly accepted. Through diagenesis fossil shells, which originally consist of calcium carbonate, are now dolomite. This process of changing a $CaCO_3$ sediment to a $CaMg(CO_3)_2$ rock is called dolomitization.

After I finished my project on Ventura turbidites, I was asked by the Shell research management to tackle the dolomitization problem. The petroleum industry is interested in the genesis of dolomite, because hydrocarbons are commonly found in this more permeable rock of a carbonate sequence. We researchers were supposed to predict the occurrence of dolomites, and knowledge of their genesis should facilitate the prediction.

I first analyzed the dolomite problem from the chemical viewpoint, and tried to define the chemistry of the solution which would react with calcium carbonate to form dolomite. Consider the dolomitization reaction

$$2\ CaCO_3 + Mg^{2+} = CaMg(CO_3)_2 + Ca^{2+}\ .$$

The equilibrium of the three phases calcite-dolomite-solution is defined by the relation:

$$(Mg^{2+}) / (Ca^{2+}) = K_{dz} \ , \tag{11.1}$$

where K_{dz} is an equilibrium constant, and is a function of temperature and pressure only. If the magnesium-to-calcium activity of a solution is greater than K_{dz}, dolomitization will take place. The foremost chemical problem was, therefore, to determine this equilibrium constant.

The conditions for three-phase equilibrium are

$$(Ca^{2+}) (CO_3^{2-}) = K_c \tag{10.23}$$

$$(Ca^{2+}) (Mg^{2+}) (CO_3^{2-})^2 = K_d \ . \tag{11.2}$$

Dividing Eq. (11.2) by Eq. (10.23), we have

$$(Mg^{2+}) / (Ca^{2+}) = K_d / K_c^2 = K_{dz} \ . \tag{11.3}$$

Since the solubility of calcite is well known, the equilibrium constant K_{dz} can be calculated if the solubility of dolomite, K_d, is determined. This is more easily said than done, because reliable experiments require approaches toward equilibrium from undersaturation and from super-saturation.

In solubility experiments, a solid phase, i.e., a crystal, can be placed in a container with a thermostat to control the experimental temperature. The solid should dissolve until the equilibrium is achieved: the ionic concentrations of a saturated solution will remain unchanged with time as long as the temperature and pressure remain unchanged. Saturation, in some instances, cannot be easily reached because dissolution usually slows down when the saturation point is approached. After many days of obtaining the same ionic concentrations in solution, an experimentalist is often uncertain whether equilibrium has been attained; the reaction rate could be so slow that changes of concentration due to continued dissolution is no longer measurable. The common practice is thus to precipitate the same solid phase from a supersaturated solution. If the final ionic concentrations of the solutions at the end of the two experiments are the same, the results indicate equilibrium and the ionic-activity product is the solubility.

Numerous experiments on dolomite solubility have been carried out, but the experimental values of apparent solubility, determined by dissolution experiments, varies from 10^{-17} to 10^{-20}. Unfortunately, dolomite cannot be synthesized at room temperature and atmospheric pressure so that the dolomite solubility determined on the basis of dissolution cannot be verified by precipitation experiments.

Nature conducts experiments of long durations. Groundwater in contact with both calcite and dolomite phases for millions of years has a better chance to achieve three-phase equilibrium than a laboratory experiment of days or months. If equilibrium is reached, or even if it is only approximated, the values of the magnesium-to-calcium concentration ratio in different groundwater samples taken from aquifers of dolomitic limestone should be more or less the same.

Shortly after I took over the project in Shell to study the chemistry of dolomite genesis, I started looking into the analytical data on groundwater chemistry. I found a very nearly constant value of the magnesium-to-calcite concentration ratio for each set of water samples. For potable water from the Florida Aquifer, the ratio indicates that the value of K_{dz} is near unity. Substituting this value into Eq. (11.3), the dolomite solubility K_d is calculated to have a value of 2×10^{-17}, which is about the same as some of the values which have been obtained from dissolution experiments. That was way back in the early 1960s, and there has not been, as far as I know, any new results to falsify this conclusion.

It should be recalled that normal seawater has a magnesium-to-calcium concentration ratio greater than 5. Taking due consideration of the different values of the activity coefficients for Ca^{2+} and Mg^{2+}, which could be calculated on the basis of the Debye-Huckle formula, the activity ratio in seawater is still not much less than 5.

We used to think that dolomitizing solutions are extraordinary rich in magnesium. Since evaporated seawater becomes enriched in Mg^{2+} after Ca^{2+} ions have been precipitated as gypsum, it was thought that such a brine should be a potent solution to form dolomite. Yet the dolomite formed in the Recent sediments of Abu Dhabi is in contact with groundwater which has a magnesium-to-calcium ratio of 2 or 3. It is also noted that aragonite or some other carbonate mineral, but hardly ever dolomite, is found in natural or artificial supersaline ponds of magnesium-rich brines. These field observations corroborate the theoretical conclusion that a very high magnesium concentration in solution is not a necessary condition for the formation of dolomite.

Seawater rich in magnesium does not precipitate dolomite because of kinetic hindrance. Miriam Kastner and her students used the expression of "surface poisoning" to describe their postulate that the high concentration of sulphate ions in seawater prevents magnesium from being ordered into a crystal lattice to form dolomite. This mineral is, however, found in marine environments where anoxic conditions prevail. Dolomite is common in diatomaceous oozes of the Gulf of California, for example, where it is formed by diagenesis in the zone of sulphate reduction.

Not all dolomite rocks are anoxic deposits. In fact most have been altered from normal marine sediments under various oxidation-reduction potentials.

After analyzing the problem for 1 or 2 years from a chemist's viewpoint, I came to the conclusion that the chemistry is not always the critical issue. In diagenetic reactions involving change of chemical composition, we have to understand the process of material transfer in and out of the system.

Dolomitization involves an addition of Mg to, and commonly also a subtraction of Ca from the system. One cubic centimeter of dolomite weighs 2.85 g and contains 0.377 g magnesium. To convert 1 cm^3 of lime sediment to 1 cm^3 dolomite with 10% porosity would require an addition of 0.34 g magnesium. This amount of magnesium has to be present in dissolved form and be transported by groundwater entering and leaving the system. The amount of the calcium subtracted from the system would depend upon the original and final porosity of the sediment.

Fast flowing groundwater, such as that in the Florida Aquifer, may contain 5 mmol or some 0.122 (=5 x 24.3 x10^{-3}) g/l of magnesium ions. For the dolomitization of 1 cm^3 lime sediment, some 3000 cm^3 of such a groundwater has to flow through the system.

How fast does the aquifer water flow? One can calculate, on the basis of Darcy's equation, or one can experiment with tracers (such as dyes or isotopes) in field measurements. Bruce Hanshaw and his colleagues used tracers and found that the Florida Aquifer flow has a velocity of about 7 m/year for traveling 137 km in central Florida (Fig. 11.1). A 700 cm/year flow rate in this dolomite of 10% porosity involves a 70 cm^3/year volume-transfer rate per unit cross-sectional area of the aquifer, or about 1/40 the amount of water needed to dolomitize 1 cm^3 of dolomite. This water delivers 8.5 x 10^{-3} g magnesium per unit cross-sectional area per year, or pushes the dolomitization front by 0.025 cm per year.

The amount of magnesium in the dolomite bed which extends for 137 km in the direction of flow is

$$0.34 \ (g/cm^3) \times 1.37 \times 10^7 \ (cm \times A \ cm^2) = 4.7 \times 10^6 \times A \ grams,$$

where A is the cross-sectional area of the dolomite aquifer. Delivering this total with the annual rate of 0.0085 x A grams, some 600 million years would have been required, even if all the dissolved magnesium in the aquifer water were used up to make this dolomite. Now the aquifer dolomite is Middle Miocene, or only about 15 million years of age. Obviously, the flow of the groundwater through the aquifer did not deliver the magnesium needed for dolomitization. In other words, the dolomite was there, and the hydrodynamic flow of the aquifer has not caused significant, if any, dolomitization.

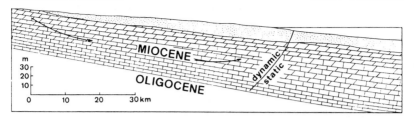

Fig. 11.1. Magnesium transfer by the Florida Aquifer flow. Magnesium needed for dolomitization has to be transported by moving water. I once thought that the Miocene dolomite which forms part of the Florida Aquifer may have acquired its magnesium from hydrodynamic movement. A simple calculation (see text) suffices to demonstrate that the rate of material transfer is much too slow for extensive dolomitization by the aquifer movement

This is but one example to illustrate the principle, that groundwater flows through carbonate aquifers, cannot be an important agent of *dolomitization*. In fact, through petrographical studies, we have found that dedolomitization, i.e., conversion of dolomite into calcite, is a common phenomenon in the host rocks of present-day groundwater.

Could more magnesium for dolomitization be delivered by faster moving groundwater? The hydraulic gradient is, for example, steep in the Prealps of Switzerland, and the permeability of the Molasse conglomerate is many orders of magnitude greater than that of the fine-grained dolomite of the Florida Aquifer. Water moves ions fast enough in and out of the conglomerate, but dolomite does not form if the chemical condition is not suitable. Groundwater in Molasse aquifers forms calcite cement, but not dolomite.

Could more magnesium for dolomitization be delivered by magnesium-rich groundwaters? Oil field brine samples from deeper formations in cratonic basins of North America have a magnesium concentration hundreds of times greater than that of the Florida Aquifer. Hydrological measurements indicate, however, that the brines are almost hydrostatic. Magnesium-rich brines set in motion by tectonic or other processes may cause dolomitization in fractures at elevated temperatures, producing the so-called hydrothermal dolomites. One of my doctoral students is studying such dolomitization phenomena in northern Italy, where Paleogene hydrodynamic circulation may have been accelerated by abnormal fluid pressure caused by overthrusting tectonics. As yet, however, we do not know enough about this mechanism to attempt a quantitative modeling of the hydrodynamics.

A popular idea on dolomitization by magnesium-rich brines is to postulate hydrodynamic circulation by seepage reflux. The word *reflux* means the return of dense brines to the ocean, after the deposition of calcium sulphate from evaporated seawater. The word *seepage* refers to a mechanism of seeping through the porous sediments between a lagoon and the open sea. Two petroleum geologists, John Emery Adams and Mary Louise Rhodes, first speculated in 1960 on this possibility as a mechanism to cause the dolomization of Permian carbonates which interbedded with evaporites (Fig. 11.2). Three colleagues at Shell Research, Ken Deffeyes, Jerry Lucia, and Peter Weyl, found the idea attractive, and formulated the "Pekelmeer" hypothesis: They noted the presence of Recent dolomite on the shore of a supersaline lagoon, locally known as Pekelmeer, on the Bonaires Island in the Caribbeans. In that lagoon the seawater has been evaporated to become sufficiently saturated with calcium sulphate to precipitate gypsum. This supersaline brine is denser so that a hydraulic head is set up to drive it back to the ocean through the sediments separating the lagoon from the open sea.

I never liked the idea because the hypothesis predicts that the carbonate in contact with the brine should be dolomitized, but the carbonate associated with gypsum in the Pekelmeer is aragonite. Dolomite is absent in Pekelmeer, but is found on the shore above the lagoon. I have also mentioned that evaporated seawater, though enriched in magnesium, does not form dolomite anywhere, due to kinetic hindrance. Yet my greatest misgiving about this fascinating idea stems from the consideration of the hydrodynamics. The Darcy-Hubbert equation states

$$q = k \ \frac{\rho g}{\eta} \cdot \left(\frac{\Delta H}{\Delta L}\right) \ .$$

The driving in this case is the density difference, which could be expressed in terms of the hydraulic gradient (dh/dl) by the consideration

$$(\rho_1 - \rho_0) \ g \ h_0 \ = \rho_0 \ g \ (h_1 - h_0) \ .$$

For a 10-m brine column which has a density of 1.13 instead of 1.03, the hydraulic potential is equivalent to a height difference of 1 m. The hydraulic gradient across a 1-km barrier separating the Pekelmeer and the open sea should thus be 10^{-3}. The permeability of fine carbonate sediment is probably in the 100-millidarcy range, or 10^{-9} cm^2. Substituting the numbers into the Darcy-Hubbert equation, we have

$$q = 10^{-9} \cdot 1.03 \cdot 981 \cdot 10^{-3} / 10^{-2} \ .$$

The quantity transferred is 10^{-7} cm^3/s or 3 $cm^3/year$ across each cm^2 of cross-sectional area. The amount of magnesium carried by this 3 cm^3 brine can at most dolomitize only 0.1 cm^3 of lime sediment. This rate is about

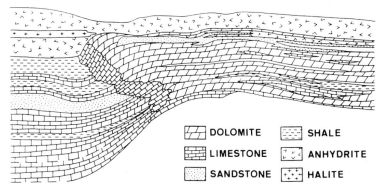

BASIN

SHELF

DOLOMITE SHALE
LIMESTONE ANHYDRITE
SANDSTONE HALITE

Fig. 11.2. A commonly observed limestone-dolomite transition. The carbonate rocks interbedded with evaporites on the platform edge are commonly dolomitized, whereas the basinal carbonates are limestones. This lateral transition first suggested to petroleum geologists that brines become enriched in magnesium (after gypsum precipitation) would react with shelf carbonate to form dolomite, while they are being refluxed back to the open sea. Calculations have shown that the rate of magnesium transfer is not fast enough for extensive dolomitization, even if such brines are reactive

four times that calculated for the Florida Aquifer water, but far from being fast enough to form dolomites many thousands of square kilometers in extent.

This Pekelmeer hypothesis fell into oblivion when one of the original co-authors went back to Bonaires Island and drilled a hole through the Pekelmeer sediments; he could find no dolomite where the supersaline brine was supposed to be refluxed back to the open sea.

When I left Shell and began teaching, I was convinced that dolomitization is a consequence of hydrodynamic circulations. But none of the modes of circulation known to me seemed capable of transporting enough magnesium to cause extensive dolomitization. My interest in this matter was rekindled in 1969 when a colleague came to give a talk on the Recent dolomite of Abu Dhabi. Dolomite is present on the sabkha, or the coastal tidal flat, everywhere above the normal spring tide. There was no Pekelmeer water to be refluxed back to the ocean. The tidal flat has practically no relief and the hydraulic gradient is almost nil. There would never be sufficient transfer of magnesium ions by lateral flow of groundwater to dolomitize the sabkha mud. Yet the dolomites found in numerous ancient carbonate sequences, several kilometers thick, are distributed over regions millions of square kilometers in extent, and those dolomites are very similar to the sabkha dolomite of Abu Dhabi. Where did the magnesium for dolomitization come from?

As I was sitting there thinking, I suddenly realized that the key word was *lateral*. We students of natural sciences grew up in temperate and humid regions, where water flows in aquifers laterally for hundreds or thousands of kilometers. We tend to forget that water in arid regions may flow vertically up or down.

One of the lessons I learned from King Hubbert was the hydrology of the Big Horn Basin in Wyoming: The basin is hydrologically closed, being surrounded on all sides by mountains. Groundwater flows through aquifers to the central depression where it has nowhere to go but up. A vertical hydraulic gradient is established because the groundwater table is so lowered by evaporative loss that its level is lower than that of the potentiometric surface of the aquifer. In other words, the water level in wells of the central Big Horn basin differs little laterally, but water in a well would rise to a considerable height above the local groundwater table. The hydraulic gradient is directed upward.

Remember that the total quantity of flow is

$$Q = q \, A \, .$$

The lateral distribution of a dolomite bed is a thousand or million times that of its vertical cross-sectional area. Giving the same linear rate q, the volume of groundwater flowing upward is a thousand or million times greater than that flowing laterally, because of the difference in cross-sectional area A. This was the key to solving the paradox, I thought, the water under the Abu Dhabi sabkha must have flowed upward: When the groundwater table is lowered by evaporation, a hydraulic gradient is set up to force the water to move vertically upward.

With the help of a student-assistant, Chris Siegenthaler, I set up a simulation experiment to demonstrate that groundwater would indeed flow upward as predicted. Evaporation lowers the hydrodynamic potential near the surface so that deeper water would move up across the pressure gradient, and evaporation seems to have the function of a pump to suck the deep groundwater up. We called, therefore, this type of hydrodynamic movement evaporative pumping.

Geology is what happened, not what could happen. The next step was, therefore, to determine whether the water was indeed flowing upward under the sabkhas where Recent dolomite had been found. The project was supported by the American Petroleum Institute, and the work was carried out during the early 1970s, mainly by two students, Jean Schneider and Judith McKenzie. The result, as shown in Fig. 11.3, is clear: The water moves upward at a rate fast enough to supply the magnesium needed to form

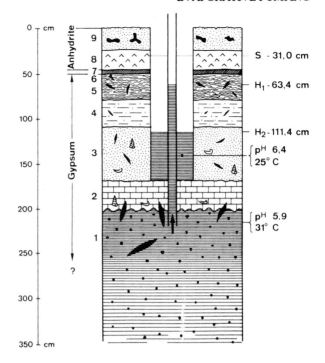

Fig. 11.3. Evaporative pumping. The potential surface of the water in the aquifer (*bed 1*, coarse sand) is higher than the groundwater table. A lithified layer (*bed 2*) separates the aquifer from the overlying Recent sabkha sediments (*beds 3, 4, 5, 6, 7*), which induce hydrodynamic movement vertically upward, called evaporative pumping. Magnesium in the subsurface water thus pumped up reacts with lime mud to form dolomite, and calcium released by the reaction reacts with sulfate in solution to form anhydrite (*beds 8, 9*)

sabkha dolomite. A model of sabkha hydrology is proposed: Rainwater in the Oman Mountains seeps down to an aquifer which transports the water laterally to the coastal region where it rises vertically due to evaporative pumping (Fig. 11.4).

Brines pumped up by evaporation should eventually be desiccated by precipitate salt, yet we found little or no rock salt in the pore space of sabkha sediments. This puzzle was finally resolved when we had the opportunity to work in Abu Dhabi after an unusual flooding of the sabkhas. Seawater driven by winds and tides to the supratidal flat seeped through the sediment and dissolved salt on its way down to recharge the aquifer (Fig.

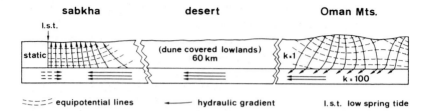

Fig. 11.4. A model of arid hydrology. Groundwater originating from rainfall recharge in the Oman mountains is transported through a permeable conduit laterally and becomes the saline groundwater in the Miocene aquifer below the sabkha. This groundwater is driven by a vertically upward-directed hydrodynamic potential, and rises to replace the near-surface evaporative loss; k = permeability in darcies

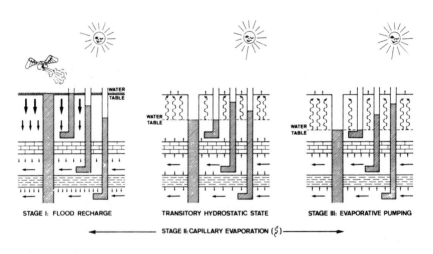

Fig. 11.5. Alternating flood recharge and evaporative pumping. The three stages of water movement after an unusual flooding of the supratidal zone are: (*I*) flood recharge, (*II*) capillary evaporation, and (*III*) evaporative pumping. The *L-shaped objects* on the right side of each diagram represent piezometric tubes buried at various depths below the surface; and the groundwater moves in the direction of potential difference. During Stage I, flood recharge water moves downward. During Stage III evaporative pumping, saline groundwater from the aquifer moves upward to replace the evaporative loss since the last flood recharge

11.5). The "traffic signal" was reversed when aquifer water was pumped upward to replace evaporative loss.

The evaporative-pumping mechanism is only operative in arid regions where evaporation far exceeds precipitation. It is thus not surprising that Recent dolomite is found only in arid regions such as South and West Australia, the Caribbeans, western North America, Persian Gulf, etc. The realization that water may move up or down has wide implications: Playa evaporites, porphyry copper ore, or even lead and zinc deposits, may have been precipitated by brines pumped up by evaporation. This excursion to groundwater hydrology illustrates a cardinal principle in natural history: Physical laws dictate. Groundwater flows according to physical laws, and diagenesis is the consequence of chemical adjustment necessitated by changes induced by physical dictates, be it temperature or pressure variations consequent upon burial, or chemical disequilibrium caused by groundwater movement.

Suggested Reading

There are literally thousands of articles written on the theme of dolomitization. I would like to recommend the two symposium volumes published by the Society of Economic Paleontologists and Mineralogists: The 1965 volume *Dolomitization and limestone diagenesis* was edited by Lloyd Pray and Ray Murray (SEPM Spec Publ 13). The article by Ken Deffeyes, Jerry Lucia, and Peter Weyl on dolomitization by brine refluxing was included in that volume; their model is an exposition of a speculative idea by J. E. Adams and M. L. Rhodes on *Dolomitization by seepage refluxion* (Am Assoc Petrol Geol Bull 44:1913-1920, 1960). The 1980 volume on *Concepts and models of dolomitization* was edited by Don Zenger, John Dunham, and Raymond Ethington. My students and I presented our model of dolomitization by *Movement of subsurface water under the sabkha, Abu Dhabi, UAE* (Judith Ann McKenzie, Kenneth J. Hsü, and Jean F. Schneider, SEPM Spec Publ 28, pp 11-30). A good summary of modern views on the chemistry of dolomitization can be found in the article on *Sedimentary rocks*, written by M. Kastner and P. A. Baker, in the McGraw Hill Encyclopedia of Science and Technology (pp 406-408, 1982).

I would like to emphasize, however, that the purpose of writing this chapter is less to review the question of dolomitization, but more to discuss the physical principles of evaporative pumping. The process is only brought into focus to illustrate a principle of hydrodynamic circulation and to evaluate its relevance to sedimentology. I have presented an analysis of a scientific problem in which I have had a lifelong interest. The evolution of my thoughts, as summarized in this chapter, is reflected by a series of my publications on this subject, namely: *Solubility of dolomite and composition of Florida groundwaters* (J Hydrol 1: 288-310, 1963); *Origin of dolomite in sedimentary sequences: a critical analysis* (Mineral Deposita 2:133-138, 1966); *Preliminary experiments on hydrodynamic movements induced by evaporation and their bearing on the dolomite problem* (with Ch. Siegenthaler,

Sedimentology 12:11-25, 1969); *Progress report on dolomitization hydrology of Abu Dhabi sabkhas, Arabian Gulf* (with J. F. Schneider, in B. H. Purser,(ed) The Persian Gulf. Berlin, Heidelberg, New York: Springer, pp 409-422, 1973).

The concept of evaporative pumping has been applied to explain the origin of evaporites by Hans Eugster and Lawrence Hardy (*Saline lakes,* in Abraham Lerman (ed) Lakes: chemistry, geology, physics. Berlin, Heidelberg, New York: Springer, pp 237-294, 1978). G. E. Smith invoked this mechanism to account for the material transfer involved in depositing the stratiform copper deposits of northern Texas (in K. H. Wolf(ed) Handbook of strata-bound and stratiform ore deposits 6:407-447, 1980). To relate evaporative pumping and sedimentology, I formulated in 1984 *A nonsteady state model for dolomite, evaporite and ore genesis* (in A. Wauschkuhn et al.(eds) Syngenesis and epigenesis in the formation of mineral deposits. Berlin,Heidelberg,New York, Tokyo: Springer, pp 275-286). My motto is clear: A knowledge of hydrodynamics is a prerequisite for an understanding of chemical diagenesis.

12   Isotopes Fractionate

Isotopes - Isotope Tracers - Paleotemperature - Paleoproductivity
Strangelove Ocean

I have promised my students not to throw technical terms around without explanations. For the sake of organization or readability, explanations may not be given when some terms are first mentioned, but discussed in a subsequent chapter. I have, for example, used the expression *isotope tracers* a few times. What are isotopes and how do they trace?

Isotopes were not discovered until 1913, because the atomic weights of chemical elements are not whole numbers as William Prout predicted in 1815. Prout's theory makes sense, because the weight of an atom is in its nuclear particles, neutrons and protons, each of which is one unit of the atom's total mass; the mass of an electron is, for all practical purposes, nil. T. W. Richards, one of the best chemical analysts of all time, proved in 1913 beyond any doubt that the atomic weights of elements are not whole numbers. But he was also upset to find that weights of one and the same element can be different. Lead produced by the radioactive decay of uranium, Richards discovered, has a different weight from that of ordinary lead. The results puzzled Richards, but a young British chemist, Frederick Soddy, proposed a brilliant solution: A chemical element is not made up of one kind of atom, but is a mixture of two or more isotopes.

The word isotope is derived from the Greek words that mean "same place". The "place" is the pigeon hole in the periodic table of chemical elements, each reserved for one chemical element which has the same number of protons and of electrons. The pigeon hole is also reserved for two or more isotopes of that element which possess different numbers of neutrons. All chlorine atoms, for example, have 17 protons in their nucleus, but one of the elements' isotopes has 18 neutrons, giving it an atomic weight of 35, while the other has 20 neutrons, giving it an atomic weight of 37. The pigeon hole for chlorine is reserved for both of the isotopes, chlorine 35 and chlorine 37, and for other isotopes which are radioactive and short-lived. Chlorine is ordinarily a mixture of the two stable isotopes. Chlorine 35 constitutes always 75.77% and chlorine 37.23% of the mixture. The atomic weight with this ratio of its two isotopes turns out to be 35.45, the inconvenient number we encountered in our chemistry exercises before computers were invented.

Isotopes are used as tracers in geology. A groundwater hydrologist, for example, may spike groundwater with a radioactive isotope, or radio-isotope, and trace the flow path through the detection of this isotope in water samples. Some radio-isotopes are, in fact, natural tracers. Carbon-14 is, for example, produced in the stratosphere by cosmic radiation. Seawater acquires its C-14 during its exchange with the atmosphere and this radio-isotope decays as it descends and circulates as a bottom current. The amount of C-14 still left in a seawater sample gives an age, which is the time elapsed since its last exchange with the atmosphere. Based upon this principle, marine geochemists, such as Wallace Broecker, could "trace" the kinematics of oceanic circulations.

Beryllium-10 has also served as a tracer of geologic processes. This radio-isotope, like C-14, is another element produced only by cosmic radiation. It is eventually precipitated as a trace element in sediments. The plate-tectonic theory postulates that oceanic deposits, which, together with ocean crust, could be subducted into the earth's mantle, partially melted, and came back out again as volcanic lavas. Recently, Be-10 has been detected in young volcanic rocks of Alaska and Japan, confirming the postulate which relates volcanism to plate subduction.

Stable isotopes which do not decay can also be used as tracers, because isotopes fractionate. This principle has provided the basis of isotope sedimentology, a rapid developing field over the last few decades. Those of you who are to specialize in sedimentology will no doubt take a course on this subject. What I shall attempt to do is to present a description of the physical principles and a few illustrations of their application to the study of sedimentary environments and sedimentary processes.

I seem to have contradicted myself when I stated that lead has different atomic weights and that chlorine has the same atomic weight of 35.45. The sameness or difference is, in fact, a matter of precision in measurements. Lead was thought to have the same atomic weight until it was proved otherwise by the superb analyst Richards. Strictly speaking, no two samples of an element need to have the same atomic weight. The proportion of atoms of various stable isotopes of an element is variable; it could deviate several parts permil or even percent from a standard value. The difference could be traced to different origins: Two lead samples may contain different proportions of lead isotopes derived from the decay of different radioactive elements. The difference in isotopic composition could also be attributed to enrichment or depletion by a natural process, which has been referred to as isotope fractionation.

Fractionation of isotopes during evaporation is a well known phenomenon: Light isotopes, like slender athletes, are more agile and proportionally more of them would escape into a gas phase when a liquid evaporates. If the system is closed, an isotope equilibrium can be established so that the isotopic compositions of an element are different in the two phases. The liquid is enriched in the heavier, and the vapor in the lighter, isotope. Of course, one could blow the vapor away, so that the isotopes can be fractionated again to make a still higher concentration of the heavier isotope in the liquid phase. Harold Urey used this principle to separate deuterium from hydrogen, and he received the Nobel Prize for his effort. To obtain his deuterium concentrate, Urey had to evaporte some 6 liters of liquid hydrogen.

Isotope fractionation takes place even if the element is not present as a pure phase. Deuterium is depleted in water vapor and is enriched in natural water bodies which have undergone a long history of evaporation. Oxygen, the other element in water, also has stable isotopes, and they are also fractionated during the evaporation of water. The most common isotope, O-16, constitutes 99.756% of oxygen atoms. Oxygen-17 and O-18 constitute 0.039 and 0.205% of oxygen. Water vapor in equilibrium with an evaporated liquid is not only depleted in deuterium but also in O-18.

Water in the open ocean has been homogenized to such an extent that its ratio of O-16 to O-18 is practically the same everywhere, and this value is used as a standard, i.e.,standard mean ocean water, or SMOW. Clouds are condensed from water vapor which has less O-18 than the ocean waters. Such a substance is said to have a negative O-18 anomaly with respect to SMOW. If O-18 in a vapour is 1 part per 1000 less than that in the SMOW, we say that this sample has a $\delta^{18}O$ value of $-1^0/oo$ SMOW. Compounds that contain more O-18 than the standard are said to have a positive O-18 anomaly.

Urey elucidated the principle of isotope fractionation during a talk in 1946 at our University. Paul Niggli, professor of mineralogy then, immediately saw an application to sedimentology. For many years geologists had difficulty in distinguishing freshwater from marine limestones. Now Urey told him that freshwater, being condensed from water vapor, has a negative oxygen anomaly. Niggli proposed to Urey that lacustrine chalk should be low in O-18 compared to marine limestone. Urey went back to Chicago and found that this is indeed the case. We have now carried the application one step further and use isotope data to distinguish the diagenetic minerals precipitated by groundwater from those precipitated in a marine environment. Isotopes have thus become tracers giving us information on depositional environments and depositional processes.

Isotope fractionation also takes place in solid-liquid reactions. Oxygen is present in water and in many solid phases. The isotopic composition of the oxygen in a solid is not the same as that in water when the two are equilibrated. The $\delta^{18}O$ value of calcite is, for example, not the same as that of a water from which it has precipitated out under equilibrium conditions. There is a differential partition of the oxygen isotopes in the solid and liquid phases: the O-18 is enriched in the solid phase. A fossil belemnite trom tne

Pee Dee Formation of South Carolina, for example, has a $\delta^{18}O$ value of plus 30.86‰ SMOW.

The enrichment of the O-18 in solid phase crystallized out of solution is temperature-dependent. At higher temperatures, more carbonate ions with heavy oxygen atoms could remain sufficiently agile to stay in solution. Calcite crystals precipitated under higher temperatures should thus be less enriched in O-18 than those formed under lower temperatures. In other words, skeletons of marine organisms precipitated under higher temperatures have a more negative O-18 anomaly than those precipitated under lower temperatures from the same SMOW. This principle permitted Urey, with the help of his associates, to devise a method in 1949 to measure the temperature of formation of solid phases during the geologic past. Incidentally, the value of SMOW is not a convenient standard to express the isotope anomalies of fossil skeletons; their values are often compared to that of Pee Dee belemnite and expressed in anomaly per mil PDB.

Quantitative analyses of isotopes are done with a mass spectrometer. A mass spectrometer, as the name suggests, separates isotopes according to their mass, or weight. This is done by "bouncing" them against a magnetic field. Lighter isotopes are "bouncier" than heavier ones, and they are therefore deflected farther from the beam streaming from an ion source. In this way ions of various masses are separated from one another and collected for counting.

The first measurement by Urey was made on a fossil skeleton from a Jurassic formation of England. The belemnite had a diameter of 2.5 cm. When a section was cut, one can see well-defined growth rings (Fig. 12.1). Samples from 24 rings were analyzed. Calibrated on the basis of experiments, the oxygen anomaly values are interpreted to be equivalent to temperature ranging from 15° to 20°C. The temperature variation suggests that the Jurassic swimmer had lived through three summers and four winters and died in the spring, age 4 years.

Urey's paleothermometry has problems, of course. Living organisms are not always law-abiding, and their vital processes might violate the rules of chemical equilibrium. Some tissues may have preference for an isotope over another and acquires more than its share of the equilibrium partition. This is called the vital effect. Fortunately, the vital effect of oxygen-isotope fractionation during skeletal growth is, in many instances, negligible in significance. Urey's tool has been refined and is more useful today than ever.

Another problem is diagenesis, which has altered fossil skeletons. The re-equilibration of isotope fractionation took place when the fossil reacts with another water with an isotopic composition different from that of SMOW, and at a temperature different from that of skeletal growth. Preservation of original skeletons, such as a fossil belemnite, is an exception rather than a rule. Urey's method was powerful, but we did not have enough samples.

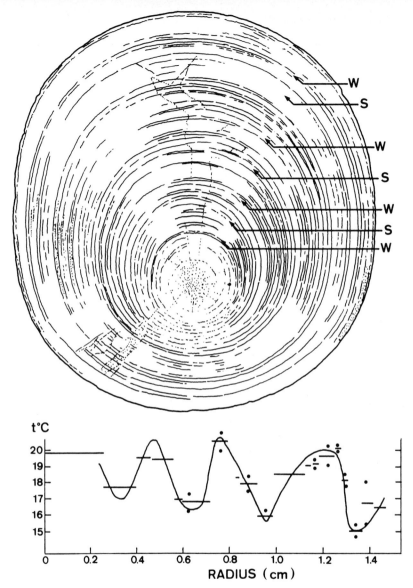

Fig. 12.1. Temperature variation in ancient ocean (after Urey et al. 1951). Making use of the principle that the oxygen-isotope composition of fossil skeletons formed in seawater is dependent upon the water temperature at the time of skeletal growth, Urey and his associates first came out with this analysis of a belemnite which indicates that the sea-surface temperature varied between 15° and 20°C during the 4 years when the organism lived. *S* and *W* indicate the parts of the skeleton which grew during the summer and winter, respectively

Major scientific progress has been made possible either through the invention of new instruments or because the opening of new horizons gives us new samples to work with. The breakthrough on the study of paleotemperatures came with deep-sea drilling. The sediments of the oceans have been, on the whole, little altered by chemical diagenesis, because the subsurface water under the seafloor is, as a rule, hydrostatic. There are exceptions, of course, and we are all excited about the implications of hot-brines spilled out of "chimneys" on the ocean floor where hydrodynamic circulation has been induced by thermal energy. There is also the hydro-dynamic movement in sediments sliced up by overthrusts, where the water is moved out of compacted sediments induced by tectonic pressure. Nevertheless, hydrostatic condition prevailed, as a rule, under the seafloor, and ocean oozes, down to 300 or 400 m depth at least, are rarely recrystallized. They are suitable specimens for paleotemperature investigations.

Urey's first trial with deep-sea sediments was a fiasco. Oozes consist of planktonic and benthic skeletal remains. Analyses of random samples containing different mixtures of the two groups gave no meaningful results. To perform a temperature analysis on oozes, warm-water floaters must be separated from cold-water bottom dwellers. A chemist cannot do that. When Cesare Emiliani, a post-doc specializing in microfossils, showed up in Chicago, he was grasped by Urey's group. The micropaleontologist knows how to tell one kind of foraminifer from another. Picking out planktonic foraminifers, Emiliani was able to use isotope analysis to show the variation of ocean-surface temperature during the Late Quaternary era. The changes in ocean climate were correlated to continental glaciations, only the depth of core penetration prevented Emiliani from getting a more complete picture of the Ice Age. The first move which led eventually to the JOIDES ocean drilling project was, in fact, Emiliani's request in 1965 for a long core to investigate the Pleistocene climate.

I was on one of the first deep-sea drilling legs to the South Atlantic. After we had completed our mission to verify the prediction of the seafloor-spreading theory, our Chief Scientist, Arthur Maxwell, reminded us that we had to drill a borehole or two for Emiliani's long core. We drilled the holes, but Emiliani was not happy with what he got. There were two problems which prevented a precise analysis of the paleotemperature record: The core was too badly disturbed, and the samples were too small to contain sufficient benthic dwellers to testify the bottom temperature.

Two developments during the 1970s opened up the possibility. Nick Shackleton of Cambridge developed a new generation of mass spectrometers. We can now analyze samples weighing 15-20 µg. Samples are no longer too small, if only a few foraminiferal tests are needed to give an analytical result. Shackleton and Kennett analyzed the cores of the Southern Oceans and were able to obtain numerical values to illustrate the decline of ocean temperatures during the Cenozoic (Fig. 12.2). The coring technique was also improved so that undisturbed cores could be taken from 300 m subbottom depth or deeper. We can take samples from intervals less than 1 cm apart to investigate ancient climate with a precision to detect changes within a 1000-year or a shorter time interval.

It was information with such precision which led me to formulate one of my more daring postulates to relate mass extinction to cometary impact at the end of the Cretaceous. Shackleton came to Zurich to present his results on DSDP Hole 384, which had been drilled into the western Atlantic bottom. His data indicated sudden increases, at the end of the Cretaceous Period, of both the surface and bottom temperatures of the oceans by 5° C. To warm up the oceans to such a degree within a matter of thousands of years seemed to require a mechanism other than normal climatic fluctuation. Was not an extraterrestrial cause at work?

It is beyond the scope of this course to discuss the theories of mass extinctions, I only mentioned the fact to illustrate a norm in science: New samples and new methods produce new data which inspire new ideas.

Carbonate rocks and minerals contain both oxygen and carbon and they are treated with acid to generate carbon dioxide to be fed into a mass spectrometer. While oxygen isotopes are being analyzed, little effort is required to determine the carbon-isotope composition of the sample. Aside from the radiocarbon C-14, the two stable carbon isotopes are C-12 and C-13. Carbon-isotope anomalies are commonly expressed in $\delta^{13}C$ per mil compared to the standard PDB.

Urey found out already in his pioneering study that carbon-isotope fractionation is not sensitively temperature-dependent. The significant mechanism of fractionation is the vital effect. Living soft tissues or cells of aquatic organisms take up a substantially lesser proportion of C-13 than that in the carbon dioxide of the surrounding water, regardless of its isotopic composition. This causes an enrichment of C-13 of the dissolved carbonate in water, and this fractionation is monitored by the skeletons of organisms living in that water.

The vital effect was confusing. I remember the conversations when I attended the Second Planktonic Conference at Kiel, 1973: There was a prevailing frustration of not knowing how to make sense of the carbon-isotope data. Plenty of analyses had been made, but the departure from isotopic equilibrium rendered uncertain the meaning of those analyses. The isotope data were usually filed and not published, because they were not understood. Only later did we realize that the cumulative vital effect as monitored by fossil skeletons, is related to the total biomass then living in a system; more biomass would produce a greater effect. The departure from equilibrium could thus serve as a measure of biomass productivity.

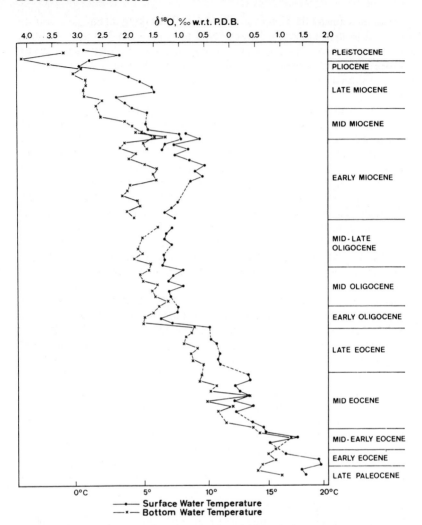

Fig. 12.2. Cenozoic temperature decline (after Shackleton and Kennett 1975). Isotope analyses of planktonic and benthic foraminifers from deep-sea drilling cores have yielded data to illustrate the general trend of global temperature decline since the Paleocene, culminating in the Quarternary glaciation. This first systematic study was followed by many such investigations on paleotemperatures over the last 15 years

Living phytoplankton and zooplankton are present only in surface waters. Because their cell tissues prefer C-12 to C-13, surface water becomes enriched in C-13 atoms wherever plankton flourishes. Bottom dwellers also shy away from C-13, but they are too thinly populated to have much effect, so that the dissolved carbonate in bottom water has a proportion of C-13 not much different from that of average ocean water. This carbon-isotope gradient has been found in open oceans (Fig. 10.3); it is also found in lakes but only in spring and summer months when planktons are blooming (Fig. 12.3).

During the late 1970s, a carbon-isotope anomaly was detected at the Cretaceous-Tertiary boundary. When I came back in 1980 from the Leg 73 drilling from the South Atlantic, I gave my samples to my associates Judith McKenzie and He Qichiang for carbon-isotope measurements. Analyzing the bulk samples, they found the negative carbon-isotope perturbation at the boundary as we expected; it has a $\delta^{13}C$ value of about -2.0^0/oo . We were still puzzling over our data when Wallace Broecker came to give an erudite exposition of the biologic pumping mechanism of carbon-isotope fractionation. He pointed out that extraordinarily high plankton productivity results in an exaggerated difference between the C-13 content in the shells of surface and bottom dwellers. What would happen if there had been no plankton production at all?

"Oh, you are asking me about the Strangelove effect", Broecker answered in high spirit when I posed the question after his talk. "An ocean without plankton would have no gradient of carbon isotopes. There would be the same carbon-isotope composition from top to bottom. The ocean would be a Strangelove ocean!"

Dr. Strangelove was the fictional character in the movie of the same name who wanted to wipe out with a nuclear holocaust all living beings on earth. Half in jest, Broecker had picked a very picturesque term, and I did not miss the chance to formalize the concept and the term in my next publication. More important is, however, the prediction: If the ocean was sparsely populated after a holocaust at the end of Cretaceous, the carbon-isotope gradient should be eliminated. The surface is no longer being depleted of C-12 atoms because of the suspension of the biologic pump. The carbonate deposited at the time should have a more negative C-13 anomaly compared to the C-13 values of the underlying and overlying sediments. This apparent enrichment of C-12 at the boundary is the carbon-isotope anomaly reported from the analysis of bulk samples, which are mainly nannoplankton skeletons. Now we could understand the negative carbon-isotope perturbation at the Cretaceous-Tertiary boundary. The next step, then, was to prove that the carbon-isotope gradient of the dissolved carbonate in seawater did not exist at the time of crisis.

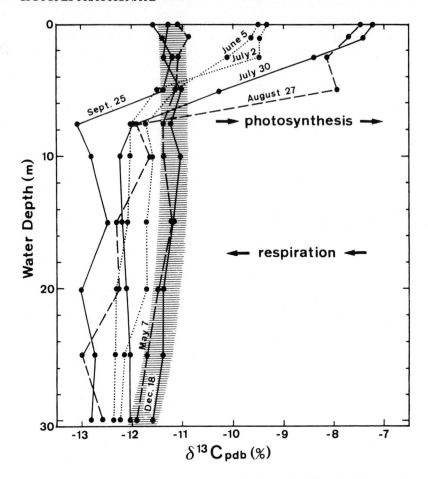

Fig. 12.3. Monthly carbon-isotope gradient of the dissolved inorganic carbon in Lake Greifen (after McKenzie 1982). The *shaded area* represents the range of $\delta^{13}C$ values found between December and May, the period of minimum productivity. The C-13 increase in surface water due to photosynthesis and the C-13 depletion in deeper waters resulting from the respiration of the sinking organic matter establish a surface-to-bottom carbon-isotope gradient during the summer months. This summer gradient is similar to the carbon-isotope gradient of the dissolved inorganic carbon in seawater (Fig. 10.3)

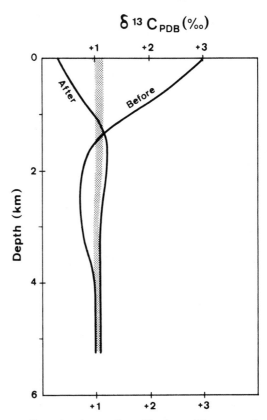

Fig. 12.4. Carbon-isotope gradient in the earliest Tertiary. The normal carbon-isotope gradient of the dissolved carbonate in seawater is characterized by a depletion of C-12 in surface water (shown by the curve *before*, meaning before the terminal Cretaceous catastrophe). The abnormal gradient (shown by the curve *after*, meaning immediately after the catastrophe) does not show this depletion when the biomass of the ocean is too small to cause the fractionation of carbon isotopes. Such a lifeless ocean is called the Strangelove Ocean

Going back to our samples we picked out planktonic and benthic foraminifers for separate analyses and compared the results (Fig. 12.4). The prediction was verified; there was no carbon-isotope gradient in the ocean after the terminal Cretaceous event.

Analyses of carbon isotopes have been informative in many other aspects of sedimentology. In analyzing the hydrocarbons in the source beds of petroleum accumulations, carbonate isotopes may give clues as to the origin of original organic substances; they carry different "signatures". Carbon isotopes are also fractionated by bacteria reactions in diagenesis, forming carbon dioxide much depleted in C-13 through sulphate reduction, and carbon dioxide much enriched in C-13 by methanogenesis reactions. Unusual carbon-isotope anomalies in sedimentary rocks are the record of such diagenetic reactions. There are books written on those studies. All I need to emphasize now is the physical principle behind all the applications, namely: The isotopes fractionate.

Suggested Reading

A most influential teacher of sedimentology today has an aversion against isotope sedimentology. He spoke contemptuously of "samples in a shoe carton" or of "numbers coming out of a black box". Many of his numerous students, who are now teachers of sedimentology in universities, share this personal prejudice. The great teacher has once pamphleteered *"In defence of field geology"*. The purpose of writing this chapter is an apology for gadgeteering sedimentology. A readable textbook is not the place to include a long list of publications on the contribution of isotope studies to sedimentology; I recommend that all students take a course in isotope geology. I shall, however, cite the references discussed in this chapter.

The use of isotopes in geology was once introduced by me in a book for general readers, *The great dying* (San Diego: Harcourt, Brace Jovanovich, 1986). In writing that introduction, I consulted the classical paper by Harold Urey, Heinz Lowenstam, Sam Epstein, and C. R. McKinney on *Measurement of paleotemperatures and temperatures of the Upper Cretaceous of England, Denmark, and the southeastern United States* (Geol Soc Am Bull Vol 62, 1951), the much cited article by Nick Shackleton and Jim Kennett *Late Cenozoic oxygen and carbon isotopic changes at DSDP Site 284* (Initial Rep Deep Sea Drilling Project 29:801-808, Washington, DC: USGov Printing Office, 1975), the new book by Wally Broecker (with T. H. Peng) on *Tracers in the Sea* (New York: Eldigo, 1982), and Judith McKenzie's study of *Carbon isotopes and productivity in the lacustrine and marine environment* (in W. Stumm(ed) Chemical Processes in Lakes. New York: Wiley, 1985). The application of the concept to interpret the isotope anomalies in the Cretaceous/Tertiary boundary sediments is presented in my article (with J. A. McKenzie) *A "Strangelove" Ocean in the earliest Tertiary* (in: Eric Sundquist and W.S. Broecker (eds) The carbon cycle and atmospheric carbon dioxide: natural variations archean to present. Am Geophys Union Geophys Monograph 32:487-492, 1985).

13   Basins Subside

Isostasy - Airy Model - Crustal Thinning - Mantle Heterogeneity Lithospheric Stresses

We have so far discussed the genesis and diagenesis of sediments and sedimentary rocks to understand the depositional processes and post-depositional alterations. A no less important subject in sedimentology is the interplay of tectonics and sedimentation, with the aim to read from sedimentary record the tectonic evolution of a region.

Thick sedimentary sequences are not uncommon, especially in folded mountain ranges. In places, sediments deposited near sea level may be several kilometers thick. The floor of sedimentary basins must have subsided in order to accomodate such thickness. Subsidence is a displacement, a physical process. What is the physical principle of basin subsidence?

James Hall invented *geosyncline* in 1859 to describe his model of subsidence. The line of maximum depression coincides with the line of maximum accumulation, in Hall's opinion, because the depression was formed by the weight of the accumulation. His contemporary, J.D. Dana, liked Hall's terminology but not his concept. Dana argued that subsidence was caused by compression and deepest depression is where thickest accumulation is possible. The chicken-and-egg argument went on for about a century: Did the chicken come first or the egg?

The idea of subsidence under sedimentary load received support from a consideration of *isostasy*, meaning equal standing. It was suggested that the earth's interior cannot sustain stress difference. The weight of sedimentary load would cause a displacement of the substratum, and the basin subsides to displace denser substratum until its weight is equal to that of the sedimentary accumulation of lesser density. The simplest model is the analogy of sea ice: When you stand on a floating ice floe, it would sink, displacing water to re-establish flotation equilibrium.

Although the term was proposed by C.E. Dutton in 1889, the concept was innovated by Airy and Pratt 30 years earlier. I do not want to repeat the oft-told story of discovery, you have no doubt learned that in your beginning geology. I will remind you, however, that the key evidence states that high mountains are compensated by a deficiency in mass of their underground.

Scientists in Airy's time thought that the earth is a molten fireball enveloped by a thin solid crust. Airy's model, or the so-called iceberg model, is formulated on the basis of such a simple idea. Crust floats on top of a denser liquid substratum like an iceberg in water. In such a flotation equilibrium, the area which is underlain by the thinnest crust has the lowest elevation and is thus a surface depression, assuming, as Airy did, that the crust has about the same density everywhere. The Royal Astronomer presented his model in a brief, but elegant exposition.

Pratt explained the deficient mass under the Himalayas by assuming an "extenuation", or physical expansion of the earth's crust under the mountains. Plateaus and mountains rise where the underlying crust expands. Pratt's model has also been described in terms of isostasy. In fact, Pratt did not have to assume flotation equilibrium above a weak substratum; he treated the problem as one of thermal expansion above a substratum which could be strong or weak. Mountains are higher where the crust is less dense because it has expanded more. We now use the term *thermal isostasy* to describe Pratt's mechanism.

When seismology was first developed around the turn of the century, it was found that the substratum under the earth's crust is not molten liquid. The substratum is a solid. It has a finite strength, and the crust should not float like an iceberg in water. We should have discarded Airy's model of isostasy. But, there came along the Yugoslavian geophysicist, Mohrovičić, who discovered a distinct surface separating the crust from a dense substratum, the mantle mantling the central core of the earth. Instead of discarding the theory of flotation equilibrium, the M-discontinuity, or Moho, was considered the undersurface of the floating crust. Assuming no lateral variations of crustal density, the compensation of surface elevation can be calculated according to the Archimedes Principle:

$$\rho_{crust} \cdot H = (\rho_{mantle} - \rho_{crust}) \cdot D' , \qquad (13.1)$$

where H is the height of the mountain above sea level and D' the thickness of the "mountain root", or that in excess of the crustal thickness of an area at sea level. Assuming a crustal density of 2.84 g/cc, and a mantle density of 3.27 g/cc, one can obtain, for example, the mountain root below a 2-km highland as 13.2 km. Assuming a 33-km thickness for the crust under the coastal plain, the crustal thickness under the mountains should be $2 + 33 + 13.2 = 46.4$ km. Similarly, the negative compensation D' of the surface depression of depth d, can be calculated according to the same principle:

$$(\rho_{crust} - \rho_{seawater}) \cdot d = (\rho_{mantle} - \rho_{crust}) \cdot D' . \qquad (13.2)$$

Assuming the same values of crustal density and crustal thickness under the coastal plain, the negative compensation under 5 km deep water should be 21.5 km. The thickness of the isostatically adjusted ocean crust should thus be $33 - 5 - 21.5 = 5.5$ km. When seismic studies in the 1940s and 1950s were able to determine the depth of Moho by explosion seismology, they found indeed a thicker crust under the mountains and a thinner crust under the ocean. The Airy model of isostasy became the paradigm.

Alfred Wegener never seemed to have suspected that Airy could be wrong. His continents drifted. Many geologists liked the idea, because the theory explained numerous facts which could not otherwise be understood, such as the Permian floras and glacial deposits of the southern continents. Many were enchanted by Gondwana, a supercontinent, which eventually broke up, the pieces drifting away to become South America, Africa, India, Australia, and Antarctica. The bulwark of resistance to Wegener came from the community of geophysicists: I remember vividly the lectures of my first American mentor, Ed Spieker; he told us of melting experiments by Goranson of the Geophysical Laboratory, who showed that basalt melts at a much higher temperature than granite. The crust cannot float, if the substratum is not molten. The geophysicists were right, and they are still right.

Nevertheless, seismologists were finding increasing evidence to support Airy's model of isostasy. Numerous studies of crustal structures were reported during *The Crust of the Earth* symposium at Columbia University in the mid-1950s. I was impressed by the accurate prediction of crustal thickness on the basis of assuming Airy isostasy. George Wollard synthesized gravity and seismic data and displayed, during an American Geophysical Union meeting, a crustal cross-section across North America, showing 50-km crust beneath the mountains, a 30-40 km crust under the continental interior, less than 30 km under the coastal plains. In sharp contrast is the thin 5 to 10 km crust under the open oceans of the Atlantic and Pacific.

The theme on the origin of sedimentary basins was one of the favorite themes of speculation in geology during the last half of the 19th and the first half of the 20th century. Theories speculated on the mechanisms which induce subsidence, but very few attempted to relate the kinematics of subsidence to changes in state as manifested by the crustal structure. Numerous mechanisms were postulated to account for the origin of "geosynclines" which could not be characterized by their crustal structure nor by the state of prevailing stresses.

I remembered the teaching of my professor in physical chemistry, who kept on telling us that the First law of Thermodynamics defines a change in state, but not the mechanism responsible for such a change. The same change can be brought about by different mechanisms, and we have to define the initial and final states before we start to theorize on mechanisms. That was the theme of my first major publication: *Isostasy and a theory for the origin of geosynclines* : Subsidence is a kinematic manifestation of a change in state, when the upper surface of the crust is depressed. What is actually changed? The change could be a change in crustal thickness, so I reasoned.

Isostatic subsidence involves crustal thinning. The initial stage is a thick crust, sufficiently thick so that its upper surface is a site of shallow marine sedimentation. Subsidence is a response when that crust becomes thinner. Thick sediments poured into a basin would further depress the basin floor to cause further subsidence. Assuming that isostatic subsidence would continue until the basin is filled with sediments up to the sea level, we could estimate the maximum thickness of sedimentary sequence in various types of sedimentary basins, $D_{sediment}$, from the relation

$$D_{sea\ level\ crust} \cdot \rho_{crust} \qquad\qquad (13.3)$$
$$= D_{sediment} \cdot \rho_{sediment} + D_{crust} \cdot \rho_{crust} + D'_{mantle} \cdot \rho_{mantle} \cdot$$

Using Eq. (13.3), and substituting values of mantle and crust density as previously mentioned, I could calculate that an ocean floor underlain by a crust 5 km thick should have an initial depth of 6.3 km. The basin would be filled up to sea level after a 13.8-km-thick sequence of sediments, which have an average density of 2.4 g/cc, is deposited; the isostatic subsidence due to sedimentary loading is thus 7.5 km. On the other hand, the isostatic subsidence under sedimentary load is only 0.15 km, if the initial crustal thickness is 32.5 and initial water depth 0.1 km.

This analysis indicates that thick sediments can be poured into deep-sea basins, as the seafloor is being depressed isostatically by sedimentary loading. Yet many "geosynclinal" sequences have been deposited on continental crust. Sedimentary loading cannot be the only cause of basin subsidence. To form a deep depression in a region underlain by continental crust requires crustal thinning, if isostatic subsidence is the cause of the origin of the basin. Therefore, one has to find the cause of subsidence through an understanding of how the crust becomes thinner under various stress environments. This message is as good today as it was in 1958. We may speak of the McKenzie model, the LePichon model, etc. , but the key question is still the why and how of crustal thinning.

Each paradigm has its Achilles' heel. Even at the height of the Airy triumph, something did not fit into the picture. When I wrote my 1958 paper, I was irritated by the work of H. E. Tuttleand M. A. Tuve. The crust under the Basin and Range, according to Eq. (13.1), should be some 40 km thick, but those two seismologists presented data to indicate a 30-km thickness there. I tried to talk my way out of the embarassment, but my arguments were arbitrary.

Sometimes I see an analogy between a paradigm in difficulty and a worn out pair of old socks. Once a hole starts to appear, your effort to mend it only leads to the appearance of another hole somewhere else, bigger and more troublesome. The misfit by Tuttle and Tuve did not go away. The work by the crustal studies group under L. C. Pakiser of U. S. Geological

Survey confirmed that the crust is "abnormally" thin under the Basin and Range. Furthermore, Airy's implicit assumption that the crust is underlain by a substratum of the same density everywhere was falling apart in the face of new evidence. Seismic velocity measurements indicate a heterogeneity of that substratum; the mantle under the Basin and Range, for example, is less dense than "normal".

The scientific community finally began to wake up to the fact that the Moho is not the undersurface of an iceberg-like feature. The upper mantle below Moho is not a molten liquid, but a solid of finite strength and the crust is not floating. Yes, there is isostasy, but the surface of isostatic compensation, i.e., of equal weight, is deeper than the Moho. We now remembered the early work by the U. S. Geodetic Survey which calculated a compensation depth of about 100 km. Joseph Barrell's old terms *lithosphere* and *asthenosphere* were exhumed to designate a lithospheric plate of finite strength, which includes the crust and upper mantle, and a weak asthenospheric mantle, respectively. We know today that continents are displaced, but they do not drift; continental crust is carried, piggyback fashion by the upper mantle, while lithospheric plates are pulled apart and pushed around.

Liberated from Airy's straightjacket that the surface elevation is dictated by crustal thickness, we are free to choose a combination of postulates to make a breakthrough. Certainly, crustal thickness is an important consideration in isostasy, but the other critical factor is lateral variation of density in the substratum, or *mantle heterogeneity*. The crust in western North America is abnormally thin in regions underlain by a hot, and therefore, lighter mantle.

Mountains are commonly underlain by a crust thicker than the surrounding one. Oceanic plateaus are underlain by a crust more than twice the normal oceanic crust thickness. In such instances, the isostatic adjustment functions according to Airy's postulate. Elevated regions such as mid-ocean ridges could, however, also be underlain by a thinner crust. The ridge crest stands high because the underlying mantle is less dense than the flanks, a corroboration of Pratt's postulate. The subsidence of the ocean floor on the flanks of a mid-ocean ridge is thus a simple process of thermal contraction. The basic principle of thermal isostasy can be expressed by Eq. (13.4), assuming a uniform expansion coefficient a for the lithosphere, and temperature δT as the "average" temperature anomaly of the lithosphere under an elevated region, the isostatic equilibrium requires

$$\rho_w \cdot d_w + \rho_l \cdot D_l$$
$$= \rho_w \cdot d_w + \rho_l (1 - \alpha \cdot \delta T) D_l + \rho_a \cdot D' , \qquad (13.4)$$

where ρ_w, ρ_l, ρ_a are the densities, and d_w, D_l, and D' are the thicknesses of the water, lithosphere, and asthenosphere column, respectively, above the level of isostatic compensation, α is the coefficient of expansion and T the temperature in Kelvin.

The relation is, of course, complicated by the fact that δT is not finite, but variable, depending upon various factors, such as the cooling history of the lithosphere in question. John Sclater and Jean Francheteau formulated an elaborate treatment to relate the seafloor bathymetry and the thermal state of the upper mantle. Current models of thermal isostasy are an extension of the Sclater concept to the subsidence of continental margins.

The controversy between Airy and Pratt is finally resolved after a century of debate. Airy's model is a special case of a general theory; the model assumes lateral homogeneity of the mantle, so that the density difference of the lithosphere is completely attributed to variations in crustal thickness. The general theory states, however, that lithospheric columns in isostatic adjustment have equal weight above a level of isostatic compensation. The difference in elevation on the surface of the earth is related to the average density difference of the lithosphere, which includes a crust of certain thickness and an upper mantle of certain density (Fig. 13.1). Viewed in this context, isostatic subsidence can be expressed as being induced by a change in the state of the lithosphere; the two variables are crustal thickness and mantle density. Changing crustal thickness is an expression of a strain related to lithospheric stress, and mantle density is a physical state related to thermal history.

Subsidence is, however, not invariably isostatic; it could also be induced by lithospheric compression, as Dana suggested. Where compressive stress prevails, isostatic compensation is hindered and surface elevation cannot be computed on the basis of isostatic consideration.

Now that the changes in state are definable, we can search for the different mechanisms. Subsidence is a displacement of the upper surface of lithosphere, and displacements are related to forces. The forces within the crust are manifested by three types of orientations of the three principal stresses: (1) Horizontal extension, (2) horizontal compression with a vertical extensional axis, and (3) horizontal compression with a horizontal extensional axis. The three types of stress orientations, as E. M. Anderson pointed out in 1951, have produced all types of faulting: the normal faults, the reverse or thrust faults, and the wrench or strike-slip faults. Since those stress states are the only states which can occur within the crust of the earth, they may also have been responsible for all types of crustal subsidence.

Extension leads to crustal thinning. If that were all that would happen, extension would induce only isostatic subsidence. Crustal extension can, however, also be induced by mantle doming, and the hot mantle material emplaced below the dome is less dense. A region may thus be uplifted during an extension, while subsidence takes place later due to crustal thinning and mantle cooling.

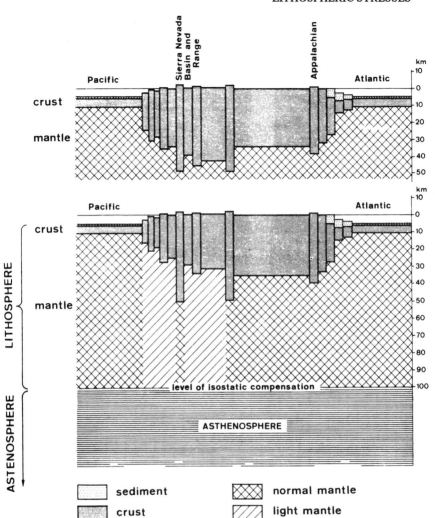

Fig. 13.1. Lithospheric isostasy. The *upper diagram* shows the Airy model of isostasy. Assuming a laterally homogeneous upper mantle of zero strength, the thickness of crust is calculated on the basis of surface elevation. The theory was falsified by seismic investigations which indicate that the crust of western North America is thinner than that predicted by the model. The lithospheric model of isostasy assumes a level of isostatic compensation considerably deeper than Moho. Furthermore, the density of the upper mantle within the lithosphere is laterally variable. This explains the apparently anomalous thin-crust of western North America in terms of lateral variation of lithospheric density

The geologic history of many regions registers such extensional phases of tectonic evolution, prior to compression and mountain building. Extension causes the breakup of a supercontinent. This rifting phase is, for example, recorded by the Permian Verrucano red beds in the Alps. The subsequent subsidence is induced by crustal thinning, combined, at a certain stage, with a cooling of the upper mantle. With the submergence of a central rift, the regions of thinner crust on both sides of the rift become passive continental margins. The thick Alpine Triassic carbonates were deposited on such a margin.

The central rift becomes eventually a site of seafloor spreading and the rifted continents are now separated by an ocean. New thin ocean crust is now formed, and thick deposits of oceanic, hemipelagic, and turbiditic sediments are accumulated in this oceanic depression. This was our Tethys Ocean, in which the Alpine Penninic sediments were laid down. The subsidence of continental margins on both sides of the Tethys is related to crustal thinning, to mantle-density change and to sedimentary loading (Fig. 13.2).

Extension changes into compression when two plates approach each other. This was the situation for the Tethys at the beginning of the Cretaceous. An oceanic trench is formed at the convergent margin, and pull-apart basins are formed. Alpine Flysch sediments, to a large part, are believed to have been dumped in such deep sea basins.

The consumption of an ocean closed the gap between continents, leading eventually to continental collision, but compression would not cease if plate displacement continues in the same direction. Mountains have been formed by collision, and the underthrusting of continental crust beneath the rising highland forms foreland basins. The Alpine Molasse are typical foreland-basin sediments.

With the revolution of earth science and the adoption of the plate-tectonic model, the old controversy on the origin of geosynclines is dead. We no longer ask the why or how, but the when and how much. Current models on the origin of sedimentary basins are quantitative evaluations of subsidence kinematics. It is beyond the scope of this beginning course to discuss the various sophisticated quantitative models, but the physical principle is simple: Subsidence is a displacement related to strain (crustal thinning) and temperature variation (density change).

Suggested Reading

The natural-history approach of learning subsidence history has led to the formulation of the concepts of geosynclines and geosynclinal cycles. The classic papers by James Hall (*Paleontology*, Geological Survey of New York, Pt 1, pp 66-96, 1859), R. D. Dana (*On some results of the earth's contraction from cooling*, Am J Sci Ser 3, 5:423-443; 6:6-14, 104-115, 162-172, 1873) and Marcel Bertrand's *Structure des Alpes francaises et récurrence de certaines*

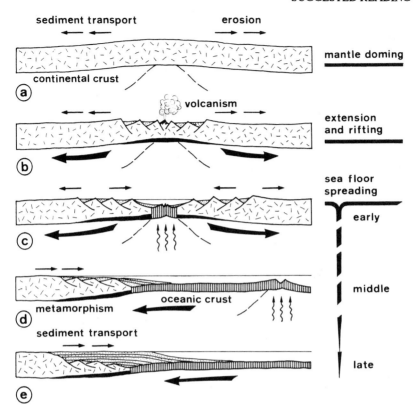

Fig. 13.2. Origin of passive continental margin. **a** Mantle doming; **b** crustal thinning and rifting; **c** emplacement of oceanic crust in the central rift; **d** seafloor spreading; **e** Sedimentary accumulation on passive continental-margins and their subsidence

faciès sédimentaires (6th Int Geol Congr, Zurich, C R pp 163-177, 1897) are brilliant syntheses by induction and make interesting reading even for students of today.

Sir G. B. Airy's *On the computations of the effect of the attraction of mountain masses disturbing the apparent astronomical latitude of stations in geodetic surveys* was published in 1855 (Phil Trans R Soc Lond 145:101-104), a few years before Archdeacon J. H. Pratt's monograph appeared (Phil Trans 149:779-796, 1859). C.E. Dutton coined the term isostasy in his paper *On some greater problems of physical geology* (Phil Soc Washington Bull 11:51-64, 1889), and Joseph Barrell discussed *The strength of the earth's crust* in 1914 (J Geol, Vol 22).

The evolution of my thoughts on the origin of sedimentary basins is manifested by a series of papers: *Isostasy and a theory for the origin of geosynclines* (Am J Sci 256:305-327, 1958), *Isostasy, crustal thinning, mantle density changes, and disappearance of ancient land masses* (A J Sci 263:97-109, 1965), *Thermal history of the Upper Mantle and its relation to crustal history in the Pacific Basin* (with Schlanger, 23rd Int Geol Congr, Prague, proceedings 1:91-105, 1968), and *Geosynclines in plate-tectonic settings* (in K. J. Hsü(ed) Mountain Building Processes. London: Academic Press, pp 3-12, 1982).

The physical-science approach to analyzing basin subsidence is exemplifed by John Sclater and Jean Francheteau's *The implications of terrestrial flow observations on current tectonic and geochemical models of the crust and Upper Mantle of the Earth* (Geophys J R Astro Soc 20:509-542, 1970), D. P. McKenzie's *Some remarks on the development of sedimentary basins* (Earth Planet Sci Lett 40:23-32, 1978), and by Xavier LePichon and Jean-Claude Sibuet's *Passive margins* (J Geophys Res 86:3708-3720, 1981).

14   Why Creativity in Geology

Geology is science. What is science? In German, we use the word *Wissenschaft*. *Wissen* is knowing, and knowing, comes from seeing (Latin *vide*). Etymologically, science is knowledge through observation. Inference through observation, or inductivism, has been a most influential methodology of science and has been held responsible for the great intellectual revolution which gave rise to modern science: Newton discovered his law of gravitation by inductively generalizing Kepler's "phenomena of planetary motion," and Kepler made his generalizations on the basis of Tycho Brahe's careful astronomical observations.

Study of the earth used to be known as *geognosy*, which was a dogma. Deductions were made on the basis of a fundamental assumption, and observations were made to harmonize with the deductions. Geology became science when Hutton wrote a theory of Earth on the basis of rationalizing, without preconceived notions, his observations. Ever since, induction from observational facts has taken precedence in geology, and deduction is considered a synonym of speculation.

The methodology of inferring general laws from particular instances has been criticized by modern philosophers, especially by Karl Popper and his followers. Inductivism has two basic assumptions, namely: Factual propositions can be derived from facts, and there can be valid inductive inferences. Popper considered both of those assumptions demonstrably false, and proposed a methodology of falsification: Only falsehood, but not truth, can be proven. Scientists of the Popperian school look for great falsifiable theories and for great negative crucial experiments. The Michelson-Morley experiment and the Eddington's eclipse experiment are the "great negative crucial experiments" which led to the formulation and general acceptance of Einstein's "great falsifiable theory."

The great attraction of Popperian deductive methodology lies in its clarity and force. Popper's definition of science has been accepted as a legal definition by Judge William Overton, when he had to decide whether the so-called creation science is science or not. Science predicts and its predictions are falsifiable. Prediction, in the form of inference from general to particular, is deduction. This form of logic has unfortunately not been sufficiently emphasized in the teaching of geology over the last hundred years; there is the incessant demand for facts and more facts, but not enough thinking of the consequences.

When I first came to Zurich, I used to tell my students:

"I come to inspire, not to inform."

Eventually, I was labeled a bad teacher. I tried to teach better, and asked my assistant what had I done wrong. I was told:

"The students come to lectures to prepare for their examinations. They would like to receive a set of written lecture notes, which they could memorize. You are considered a bad teacher, because students cannot take voluminous notes during your lecturing. They like your stories, but they feel that they are not learning anything."

Yes, Stokes' law is a corollary of Newton's First Law of Motion, which they had learned during their middle-school days, so they thought. The Chezy equation is an empirical formula in hydraulic engineering; it has little relevance to earth science. Thermodynamics is not a favorite pastime for geology students; they would have studied physics or chemistry if they had been interested in such purely theoretical stuff. My young friend finally ended his discourse with advice:

"The problem is that you try to inspire creativity, but the students want to be informed."

What is creativity?

Creativity in arts has been illustrated by the sculpturing of Michelangelo, the paintings of Van Gogh, the compositions of Bach, or the dramas of Shakespeare. Michelangelo achieves his perfection in *Pieta* by chipping off all the superfluous pieces of marble. Shakespeare manifests his creative genius, when he disregarded superfluous words and chose only those and in such order as he wrote *King Lear* or *Othello* .

Science is searching for truth. We know truth exists because we can recognize falsehood. Creativity in science is to recognize the falsehood. False ideas in science are the pieces of marble chipped away by Michelangelo in his creation of *Pieta*; they are the words not used by Shakespeare in his writing of *Othello* .

What is so creative about *David* or about *Othello* ? They are inferences from general to particular. *Pieta* is a particular glorification of an ideal, and *Othello* is a particular tragedy of fraility in human character. The inference, the imaginative deduction, is creativity.

Why creativity in science?

In this question, my colleagues thought they had found the answer as to why I had been such a poor teacher. They would tell me:

"You must not forget that E. T. H. is not M. I. T. or Cal Tech. We are not a university for the elite only. We have to accept any student who has received a "maturity" certificate on leaving middle school. Many of them come to a technical university to learn a profession. Einstein was a singular exception at E. T. H. We cannot try to make all of them Einsteins."

I have listened to their arguments and have changed my teaching style somewhat. I tried to be more informative. I tried to teach more the natural-history approach in sedimentology. The first lecture was no longer Newton's First Law of Motion, but the geologic significance of mineral composition. I take cognizance of the fact that the duty of geological engineers is to make observations, gather facts, and to write an informative report. It falls beyond their competence to indulge in all kinds of why's and if's. Inductivism in science and narrow specialization seems to make good practioners in a profession which has no need for creative thinking.

This type of mentality has led to disasters. Nature is capricious, and we need imaginative geological engineers to understand her seemingly unpredictable mood and to appreciate the grave consequences of her tantrums. It has happened all too often, when a colleague went into a depression because a fallen block from a glacier killed more than 100 workers on his engineering project, or another committed suicide when a dam which he had helped design broke and the flood drowned thousands of people. A team of observers of active volcanoes arrived in Peru after a mudslide buried a village, or a weather bureau neglected to foresee the danger of storm-driven waves to thousands of villagers in western Louisiana. Why did the seismologists who achieved worldwide fame in earthquake prediction fail to give the slightest warning to the 270000 victims of the Tangshan catastrophe? Why did the geologists whose descriptions of the Alpine tectonics have become a classic fail to forecast the drowning of poor diggers of the Simplon Tunnel. I do not know the history of those events well enough, but I know at least two tragedies which could have been avoided if the scientists and engineers in charge had learned creativity in science.

Take the Vaiont catastrophe for example, floodwater spilled over a dam killing 5000 persons. Geologists and engineers were aware of the slope instability above the reservoir lake behind the Vaiont dam. They had been monitoring the earth's movement for several years. Based upon inductive reasoning that failure would come soon when the creep rate accelerates, scientists were predicting the imminent breakup of the big sliding mass. Even the volume of the mass was estimated. None made, however, an inference with this conclusion: What would happen if this mass was to fall into the lake. The deduction is simple: The slide would move instantaneously under gravity to a depression, i.e., the lake basin, and would displace the water of the lake. The water would spill over the dam, rush down the canyon and drown the population of a whole town at the mouth of the canyon. No such inference was made because scientists and engineers have been trained to stick to their job: Let's do good science, have facts, measure our creep rate. It does not seem scientific to make wild speculations. They do not understand that deduction is not speculative: deduction is good science. Consequence: tragedy occurred.

I have been impressed by another case of stupid mediocrity, a tunnel tragedy. A geologist was asked to make a geological investigation in connection with a tunnel construction in the Alps. Routinely he made field observations, laboratory studies, mapped the area, drew a cross-section. He described the geologic structure, which is a folded thrust: Triassic dolomite is overthrust above Jurassic carbonate rocks. The thrust plane is a gypsum layer, and that is a very common situation in the Alps. Most important to him seemed to be the fact that the carbonate rocks are not friable, and there should be no big problem boring a tunnel through it, so the young man concluded. Workers were sent to dig a shaft horizontally from the Jurassic to the Triassic. They dug through the thrust, and then the gypsum layer. Water suddenly rushed into the tunnel and drowned all the workers. The case was finally declared an accident, because the flooding was not foreseen. It was not foreseen, because the geologist did not use his deductive reasoning: Permeable dolomite above an impermeable gypsum in a syncline is an ideal water-bearing rock with a perched groundwater table. But that was a hydrological problem. Our geologist was trained in stratigraphy and geological structure, it was not up to him to "speculate" what would happen if you drill through an aquiclude in such a situation. Consequence: an accident occurred.

Not all the mistakes were made by engineering geologists. I worked in an industry and heard many tales, too. Is it purely luck that the largest gold, silver, and copper deposits of the world were found by George Hearst, a person who never went to university? Is it only coincidence that the largest oil fields of Libya were not discovered by the major oil companies? We do not know the answer to the first question, but someone gave me the answer to the second one, when I asked:

"You minors seem to have been so successful in Libya that you cannot miss. No matter where you drilled, you made a discovery. Why did Company X, with all their big staff and heavy investment fail to make a single strike?"

"The Company X discovered many oil fields in the Paleozoic of North Africa. Their people thought that oil had to be found in the Paleozoic. So they came in early and took all the "good" acreage. When we got into Libya, we had to drill in the Tertiary, because that was the only place left for us. Their arrogance has costed them billions!"

Those arrogant geologists never read Popper, or else they would have realized that an inference based upon 99 observations could be falsified by the hundredth.

Lack of success in exploration is less traumatic an experience than malpractices resulting in loss of life. Yet activities by exploration geologists are not necessarily trivial, and their failure may have grave consequences influencing the course of history. My favorite story is the inability of the Japanese geologists to discover the Daqing Oil Field in Manchuria:

Northeast China came under Japanese influence after the 1905 Russo-

Japanese War, and Japan occupied Manchuria in 1931. For four decades until 1945, geologists searched everywhere for oil in those provinces. They did much work in the Songliao Basin, describing the sedimentary sequence of lacustrine deposition, and dating their geologic age by micropaleontology. Yet they did not drill a single test hole in the basin although a big anticline was present. They ruled out the potential of Songliao because big oil fields had been disovered only in marine deposits. But this inductive generalization was wrong. Chinese scientists came to Songliao in the late 1950s. They drilled nine shallow test holes, and six of those at Daqing struck oil. Now this giant oil field accounts for more than one-half of the total Chinese production.

I remember the critical days before Pearl Harbor, when the Japanese military were debating whether they should go north or south. I am almost certain that Japan would not have started a war with the United States if they had discovered Daqing. The Japanese needed petroleum products for their war-making, and the Indonesian oil was the only "available" source for them. They attacked Pearl Harbor because they hoped to eliminate, at least temporarily, effective military intervention from the west when they were invading the south. If they had had Daqing, they would not have had to fight America for their oil. Instead, they would have to take defensive measures against potential invaders from Siberia. They might even have helped Hitler and defeated the Soviet Union, if they had not made war against the United States. We might be living in a totally different world today if the Japanese had found Daqing.

In this speculative mood, I shall conclude my opus with an epitaph:
**Deduction is science, not speculation,
and there is no substitute for creative thinking.**

Suggested Reading

Harvey Blatt in his textbook on *Sedimentary petrology* (San Francisco: Freeman, 1982) included a chapter on the practice of sedimentology. I applaud his appreciation of the need, but a cookbook approach is no solution to scientific methodology. Scientific endeavors, like all other intellectual activities, require an analytical mind. Before a method can be devised, questions have to be asked: What is science? What are we trying to achieve? This brings us to the realm of philosophy of science.

Readers preparing for examinations will have neither the interest nor the time to delve into philosophical questions. Those who are taking a vacation on the seashore or convalescing from a long illness may wish to read for pleasure *Scientific revolutions* edited by Ian Hacking (New York: Oxford Univ Press, 1981); included in this slim volume are gems by T. S. Kuhn, Dudley Shapere, Karl Popper, Imre Lakatos, Ian Hacking, Larry Laudon, and Paul Feyerabend. A person might be sufficiently inspired to make a study of *The*

205

structure of scientific revolutions by Kuhn (Chicago: Chicago Univ Press, 2nd ed, 1970), *Conjectures and refutations* by Popper, and/or *Falsification and the methodology of research programs* by Lakatos.

Appendix I

Symbols used in this text and their dimensions

(Not included in this list are conventional mathematical symbols and notations for chemical elements)

A	area, L^2
(A)	concentration or activity of chemical component A
a	dimensionless number in mass-action law
a	acceleration, $M\,L\,t^{-2}$
$\underline{a}$	chemical activity
(B)	concentration or activity of chemical component B
b	dimensionless number in mass-action law
C	Chezy's coefficient
C_f	resistance coefficient, $L^{1/2}\,t^{-1}$
C_s	a dimensionless constant
(C)	concentration or activity of chemical component C, dimensionless ($M\,M^{-1}$)
$\ddot{C}$	concentration of solute in pore water, $M\,L^{-3}$
C	compaction index, dimensionless
c	dimensionless number in mass-action law
c	coefficient of diffusion, $L^2\,t^{-1}$
c	subscript, referring to a critical value
D	a linear dimension, diameter or thickness, L
(D)	concentration or activity of chemical component D
d	dimensionless number in mass-action law
d	a linear dimension, commonly referred to depth, L
E	energy or energy content, $M\,L^2\,t^{-2}$
F	force, $M\,L\,t^{-2}$
F_g	gravity force, $M\,L\,t^{-2}$
F_i	inertial force, $M\,Lt^{-2}$
F_r	resistance force, $M\,L\,t^{-2}$
F_v	viscous force, $M\,L\,t^{-2}$
Fr	Froude number, dimensionless
f	function of
f	subscript, referring to fluid

g	gravitational acceleration, M L t^{-2}
H	a linear dimension, commonly referred to height, L
K	transmissibility, L t^{-1}
K$_{subscript}$	scale factor in model theory
k	permeability, L^2
K.E.	kinetic energy, M L^2 t^{-2}
L	a linear dimension, L
M	mass, M
m	mole number of component i in solution, M
m	subscript referring to mass in model theory
o	subscript in model theory for original
P	power, M L^2 t^{-3}
P$_g$	power of gravity force, M L^2 t^{-3}
P$_r$	power of resistance, M L^2 t^{-3}
P$_s$	power of grain settling, M L^2 t^{-3}
P.E.	potential energy, M L^2 t^{-2}
p	pressure, M L^{-1} t^{-2}
Q	heat, M L^2 t^{-2}
Q	volume flow rate, L^3 t^{-1}
q	linear flow rate, L t^{-1}
q	ionic diffusion rate, M L^{-2} t^{-1}
R	radius of capillary tube
r	distance from wall of capillary tube
R	gas constant, M L^2 t^{-2} (Kelvin)$^{-1}$
Re	Reynolds number, dimensionless
Re*	boundary Reynolds number, dimensionless
Ri	Richardson number, dimensionless
S	entropy, M L^2 t^{-2} (Kelvin)$^{-1}$
s	distance, L
s	slope, dimensionless
s	subscript, referring to solid
T	temperature, Kelvin
t	time, t
U	velocity, L t^{-1}
u	velocity, L t^{-1}
u$_s$	settling velocity, L t^{-1}
u$_f$	fluid velocity, L t^{-1}
u*	shear velocity, L t^{-1}
V	volume, L^3

W	work, $M L^2 t^{-2}$
w	a linear dimension, commonly referred to width, L
X	a linear dimension, L
x	a linear dimension, L
y	a linear dimension, L
Z	a dimensionless number of indefinite value
z	a linear dimension, L
α	coefficient of expansion, dimensionless
δ	per mil deviation in isotope value from a standard
η	viscosity, $M L^{-1} t^{-1}$
θ	an angle, dimensionless
ρ	density, $M L^{-3}$
ρ_f	density of fluid, $M L^{-3}$
ρ_s	density of solid, $M L^{-3}$
σ	stress, $M L^{-1} t^{-2}$
τ	shearing stress, $M L^{-1} t^{-2}$
μ	coefficient of friction, dimensionless
$\mu_{subscript}$	chemical potential, $L^2 t^{-2}$
ν	kinematic viscosity

Appendix II

Quantitative relations in the physical principles of sedimentology

Stoke's law (settling velocity):

$$u = \frac{1}{18} \frac{(\rho_s - \rho_f)\, g\, D^2}{\eta} \tag{1.1}$$

Fluid resistance

$$F_r = C_f \frac{\rho\, u^2}{2}\, A \tag{3.10}$$

Reynolds number (criterion of turbulence)

$$\mathbf{Re} = \frac{\rho\, u\, D}{\eta} = \text{dimensionless number} \; \frac{\text{inertial force}}{\text{viscous force}} \tag{3.13}$$

Chezy's equation (stream-flow velocity)

$$u = C \sqrt{d \cdot s} \tag{4.1}$$

Chezy-Kuenen equation (density-flow velocity)

$$u = C \sqrt{\Delta \rho_c \cdot d \cdot s} \tag{4.2}$$

Shields criterion (shear velocity to initial grain movement)

$$u^* = 0.06 \sqrt{(\rho_s - \rho_f)\, g \cdot D} \tag{4.6}$$

APPENDIX II

Heim-Müller equation (rockfall acceleration)

$$a = g \ (\sin \theta - \mu \cos \theta)$$ (5.5)

Bagnold's criterion (auto-suspension)

$$\sin \theta = \left(\frac{1}{2} \frac{C_f}{R_i} + \frac{u_s}{u_f} \right)$$ (6.7)

Keulegan's law (velocity of saline head)

$$u = 0.71 \sqrt{\frac{\Delta \rho}{\rho} g \ d}$$ (6.10)

Richardson number (criterion for mixing)

$$R_i = \frac{\frac{\Delta \rho}{\rho} g \ d}{d \ t^2} = \text{dimensionless number} \ \frac{\text{gravity force}}{\text{inertial force}}$$

Froude number (criterion for flow regime)

$$\mathbf{Fr} = \sqrt{\frac{u^2}{g \ D}} = \frac{u}{\sqrt{g \ D}}$$ (7.17)

Densiometric Froude number (criterion for flow regime)

$$\mathbf{Fr_d} = \frac{u}{\sqrt{\frac{\Delta \rho_c}{\rho} g \ D}}$$

Bernoulli theorem

$$\frac{p}{\rho} + g \ H + \frac{u}{2} = \text{constant}$$ (8.2)

Darcy-Weisbach equation (resistance to fluid flow in pipe)

$$\tau_r = \frac{f}{4} \cdot \frac{\rho u^2}{2} \tag{8.12}$$

Darcy's equation (flow velocity through porous media)

$$Q/A = K \, (\Delta H)/(\Delta L) \tag{9.1}$$

Poiseuille's law (velocity of viscous flow through a tube)

$$u = \frac{\Delta p}{4\eta \Delta L} \left(R^2 - r^2\right) \tag{9.5}$$

Darcy-Hubbert equation (flow velocity through porous media)

$$q = k \, \frac{\rho g}{\eta} \cdot \left(\frac{\Delta H}{\Delta L}\right) \tag{9.12}$$

Kinetics of ionic diffusion

$$q = c \left(\frac{d\dot{C}}{dx}\right)$$

Mass action law

$$(C^c) \, (D^d) \, / \, (A^a) \, (B^b) = K \tag{10.1}$$

Gibbs criterion (chemical equilibrium)

$$T'dS' - p'dV' + \mu_1' \, dm_1' + \mu_2' dm_2' + ... + \mu_n' dm_n' \tag{10.4}$$
$$+ \, T''dS'' - p''dV'' + \mu_1'' dm_1'' + \mu_2'' dm_2'' + ... + \mu_n'' dm_n''$$
$$+ \, T'''dS''' - p'''dV''' + \mu_1''' dm_1''' + \mu_2''' dm_2''' + ... + \mu_n''' dm_n''' = 0$$

APPENDIX II

A solution of Eq. (10.4) is

$$T' = T'' = T'''$$
$$p' = p'' = p'''$$
$$\mu_1' = \mu_1'' = \mu_1'''$$
$$\mu_2' = \mu_2'' = \mu_2'''$$
$$\mu_n' = \mu_n'' = \mu_n''' \qquad (10.5)$$

First Law of Thermodynamics

$$dE = dW - dQ \qquad (10.10)$$

Second Law of Thermodynamics

$$dS = \frac{dQ}{T} \qquad (10.14)$$

Lewis definition of chemical activity

$$\mu = \mu_0 + RT \ln \underline{a} \qquad (10.20)$$
$$\lim_{c \to o} (\underline{a}/c) = 1$$

Airy model of isostasy

$$\rho_{crust} \cdot H = (\rho_{mantle} - \rho_{crust}) \cdot D' \qquad (13.1)$$

$$(\rho_{crust} - \rho_{seawater}) \cdot d = (\rho_{mantle} - \rho_{crust}) \cdot D' \qquad (13.2)$$

$$D_{sea\ level\ crust} \cdot \rho_{crust} \qquad (13.3)$$
$$= D_{sediment} \cdot \rho_{sediment} + D_{crust} \cdot \rho_{crust} + D'_{mantle} \cdot \rho_{mantle}$$

Pratt model of isostasy

$$\rho_w \times d_w + \rho_1 \cdot D_1 = \rho_w \cdot d_w + \rho_1 (1 - \alpha \cdot \delta T) D_1 + \rho_a \cdot D' \qquad (13.4)$$

References

Adams H (1936) Activity and related thermodynamic quantities; their definition and variation with temperature and pressure. Chem Review 19:1-26

Adams JE, Rhodes ML (1960) Dolomitization by seepage refluxion. AAPG Bull 44:1913-1920

Airy GB (1855) On the computations of the effect of the attraction of mountain masses disturbing the apparent astronomical latitude of stations in geodetic surveys. Royal Soc London, Phil Trans 145:101-104

Allen JRL (1970) Physical processes of sedimentation. Allen and Unwin, London, 46 pp

Bagnold RA (1962) Auto-suspension of transported sediments. Proc Royal Soc series A 265:315-319

Bailey EB (1936) Sedimentation in relation to tectonics. Geol Soc Am Bull 47:1713-1726

Barrell J (1914) The strength of the earth's crust. J Geol, 22:28-48, 145-165, 209-236, 289-314, 441-468, 537-555, 655-683, 729-741

Beneo F(1955) Les résultats des études pour la recherche pétrolifère en Sicilie. Proc 4th World Petrol Congr, Sec 1:121-122

Berger WH, Winterer EL (1974) Plate stratigraphy and the fluctuating carbonate line. In: Jenkins H, Hsü KJ (eds) Pelagic sediments. Int Assoc Sedimentology, Spec Publ 1:11-48

Bernard H et.al.(1970) Recent sediments of southern Texas. Guidebook 11, Bureau of Economic Geology of Texas, 83 pp

Bertrand M (1897) Structure des Alpes francaises et récurrence de certaines faciès sédimentaires. 6th Int Geol Congress, Zurich, compt rend pp 163-177

Blatt H (1982) Sedimentary petrology. Freeman, San Francisco, 564 pp

Bouma A (1962) Sedimentology of some flysch deposits. Elsevier, Amsterdam, 168 pp

Broecker W, Peng TH (1982) Tracers in the sea. Eldigio Press, New York, 690 pp

Buss E, Heim A (1881) Der Bergsturz von Elm. Wurster & Cie, Zurich, 163 pp

Clarke J (1988) Gravel waves deposits. Diss: Dalhousie University

Clifton HE (ed) Sedimentologic consequences of convulsive geologic events Geol Soc Am Spec Pub (in press)

Dana RD (1873) On some results of the earth's contraction from cooling. Am J Sci Ser 3, 5:423-443, 6:6-14, 104-115, 162-172

Darcy H (1856) Les fontaines publiques de la ville de Dijon. Victor Dalmont, Paris, pp 590-594

Deffeyes KS, Lucia FJ,Weyl.PK (1965) Dolomitization of recent and Plio-Pleistocene Sediments ba marine evaporate waters on Bonaire, Netherlands Antilles. SEPM Spec Publ 13:71-88

REFERENCES

Dutton CE (1889) On some greater problems of physical geology. Phil Soc Washington, Bull 11:51-64

Eugster H, Hardy L (1978) Saline Lakes. In: Lerman A (1978) Lakes - chemistry, geology, physics. Springer, Berlin, Heidelberg, New York pp 237-294

Flores G (1955) Discussion of a paper by F. Beneo: Les résultats des études pour la recherche pétrolifêre en Sicilie. Proc 4th World Petr Congr, Sec 1:121-122

Gibbs JW (1875-1878) On the equilibrium of heterogeneous substances. Reprint (1961) The scientific papers of J. Willard Gibbs, 1:55-349, 419-425.Dover Publications, N.Y.

Ginsburg R (1956) Environmental relationships of grain size and constituent particles in some South Florida carbonate sediments. Bull Am Assoc Petr Geol 40:2384-2427

Ginsburg R (1973)(ed) Evolving concepts of sedimentology. Pettijohn Festschrift. Johns Hopkins Univ Studies in Geology, no 21, 191 pp

Hacking I (ed)(1981) Scientific revolutions. Oxford University Press, New York, 180 pp

Hall J (1859) Paleontology. Geological Survey of New York, Pt. 1:66-96

Hampton HA (1972) The rôle of subaqueous debris flows in generating turbidity currents. J Sed Petr 42:775-793

Heezen B, Hollister CD (1971) The face of the deep. Oxford University Press, N.Y., 659 pp

Heim A (1882) Der Bergsturz von Elm. Deutsche Geol Ges Zeitschrift 34:74-115

Heim A (1932) Bergsturz und Menschenleben. Fretz und Wasmuth, Zurich, 218 pp

Hjulström F (1939) Transportation of detritus by moving water. Trask PD (ed) Recent marine sediments. Am Assoc Petr Geol, Tulsa, pp 5-41

Hsü KJ (1958) Isostasy and a theory for the origin of geosynclines Am J Sci 256:305-327

Hsü KJ (1963) Solubility of dolomite and composition of Florida ground waters. J Hydrol 1:288-310

Hsü KJ (1964) Cross-laminations in graded bed sequences. J Sed Petr 34:379-388

Hsü KJ (1965) Isostasy, crustal thinning, mantle density changes, and disappearance of ancient land masses . Am J Sci 263:97-109

Hsü KJ (1966) Origin of dolomite in sedimentary sequences: a critical analysis. Mineralium Deposita 2:133-138

Hsü KJ (1967) Chemistry of dolomite formation. In: Chilingar GV Bissell HJ, Fairbridge RW.(eds) Carbonate Rocks. Developments in sedimentology 9B:169-191, Elsevier, Amsterdam

Hsü KJ (1975) Catastrophic debris streams (sturzstroms) generated by rockfalls. Geol Soc Am Bull 86:128-140

Hsü KJ (1976) Paleoceanography of the Mesozoic Alpine Tethys. Geol Soc Amer Spec Pap 170:27-36

Hsü KJ (1977) Ventura basin. AAPG Bull 61:137-168, 169-191

Hsü KJ (1978) Albert Heim: Observations on landslides and relevance to modern interpretations. In: Voight B (ed) Rockslides and avalanches. Elsevier, Amsterdam, 833 pp

Hsü KJ (1979) Ventura basin. AAPG Bull 64:1034-1051

Hsü KJ (1980)Texture and mineralogy of the Recent sands of the Gulf Coast. J Sed Petr 30: 380-403

Hsü KJ (1982) Geosynclines in plate-tectonic settings. In Hsü KJ (ed) Mountain building processes. Academic Press, London, pp 3-12

Hsü KJ (1983) When the Mediterranean was a desert. Princeton University Press, Princeton NJ, 197 pp

Hsü KJ (1984) A nonsteady state model for dolomite, evaporite and ore genesis. In: Wauschkuhn A, Kluth C, Zimmermann RA.(eds) Syngenesis and epigenesis in the formation of mineral deposits. Springer, Berlin, Heidelberg, New York, pp 275-286

Hsü KJ (1986)The great dying. Harcourt, Brace Jovanovich, San Diego, 292 pp

Hsü KJ (1988) Melange and melange tectonics of Taiwan. Proc Geol Soc China 31:87-92

Hsü KJ, Kelts K (1985) Swiss lakes as a geological laboratory. Naturwissenschaft 72:315-321

Hsü KJ, McKenzie JA (1985) A "Strangelove" ocean in the earliest Tertiary. In: Sundquist E, Broecker, WS (eds) The carbon cycle and atmospheric carbon dioxide: natural variations Archean to Present. Am Geophysical Union, Geophysical Monograph 32:487-492

Hsü KJ, Schlanger S (1968) Thermal history of the Upper Mantle and its relation to crustal history in the Pacific Basin. 23rd Int Geol Congr, Prague, Proc 1:91-105

Hsü KJ, Schneider JF (1973) Progress report on dolomitization hydrology of Abu Dhabi sabkhas, Arabian Gulf. In Purser BH (ed) The Persian Gulf. Springer, Berlin, Heidelberg, New York, pp 409-422

Hsü KJ, Siegenthaler C(1969) Preliminary experiments on hydrodynamic movements induced by evaporation and their bearing on the dolomite problem. Sedimentology 12:11-25

Kastner M, Baker PA (1982) Sedimentary rocks. McGraw Hill Encyclopedia of Science and Technology pp 406-408

Kelts K (1978) Geological and sedimentary evolution of Lakes Zürich and Zug, Switzerland. Diss ETH 6146, Zürich, 250 pp

Kersey D, Hsü KJ (1976) Experimental investigation of the energy relations of density-current flows. Sedimentology 23:761-789

Keulegan GH (1946) First and second progress reports on project 48. Natl Bureau of Standards Rep pp 1-17

Keulegan GH (1957/8) 12th and 13th Progress reports on model laws for density currents. Natl Bureau of Standards Rep 5168:1-28, 5831:1-18

Hubbert MK (1937) Theory of scale models as applied to studies of geologic structures. Geol Soc Am Bull 48:1459-1520

Hubbert MK (1938) The place of geophysics in a department of geology. AIME Tech Publ, No 945

Hubbert MK (1940) Theory of ground-water motion. J Geol 48:785-944

Hubbert MK (1949) Report on the Committee on Geologic Education of the Geological Society of America. Interim Proc Geol Soc Am

Hubbert MK (1962) Presidential address of 1962. Geol Soc Am Bull, 74:365-378

REFERENCES

Kuenen Ph (1952) Estimated size of the Grand Banks turbidity current. Am J Sci 250:874-884

Kuenen Ph, Migliorini CI (1950) Turbidity currents as a cause of graded bedding. J Geol 58:91-127

Kuhn TS (1962) The structure of scientific revolutions. Chicago University Press, 172 pp

Lacombe H, Tchernia P (1971) Caractères hydrologiques et circulation des eaux en Méditerranée. In: Stanley DJ (ed) The Mediterranean Sea. Dowden, Hutchinson & Ross, Stroudsbourg,Pa pp 25-36

Lakatos I (1970) Falsification and the methodology of research programs. In: Lakatos I, Musgrave A (eds) Criticism and the growth of knowledge. Cambridge Univ Press, Cambridge

Lambert A, Hsü KJ (1979) Varve-like sediments of the Walensee, Switzerland. In Schlüchter Ch (ed) Moraines and varves. Balkema, Rotterdam, pp 287-294

Lambert A, Giovanoli F (1988) Limnology and oceanography 33:458-468

Langhaar H L (1951) Dimensional analysis and theory of models. Wiley, New York, 166 pp

Leeder MR (1982) Sedimentology. Allen and Unwin, London 75 pp

Leopold L, Wolman G, Miller J(1964) Fluvial processes in geomorphology. W. H. Freeman, San Francisco, 522 pp

LePichon X, Sibuet J-C (1981) Passive margins. J Geophys Res 86:3708-3720

Lüthi S (1978) Zur Mechank der Turbiditätsströme. Diss ETH 6258, Zürich, 160 pp

Mackie W (1896) The sands and sandstones of eastern Moray. Trans Edinburgh Geol Soc 7: 148-172

Malinverno A, Ryan WBF (1985) Large avalanche scars on the continental margin. In: Clifton HE (ed) Sedimentologic consequences of convulsive geologic events. Geol Soc Am Spec Publ (in press)

Mandelbrot BB (1982) The fractal geometry of nature. Freeman, San Francisco, 460 pp

McKenzie JA (1982) Carbon-13 cycle in Lake Greifen: a model for restrictive ocean basins. In: Schlanger SO, Cita MB (eds) Nature and origin of Cretaceous carbon-rich facies. Academic Press, New York, pp 197-207

McKenzie JA (1985) Carbon isotopes and productivity in the lacustrine and marine environment. In W. Stumm (ed) Chemical processes in lakes. Wiley,New York, pp 99-117

McKenzie JA, Hsü KJ, Schneider JF (1980) Movement of subsurface water under the sabkha, Abu Dhabi, UAE. SEPM Spec Publ 28:ll-30

McKenzieDP (1978) Some remarks on the development of sedimentary basins. Earth and Planetary Sci Lett 40:23-32

Menard W (1964) Marine geology of the Pacific. McGraw Hill, N.Y., 271 pp

Middleton GV (1964) Experiments on density and turbidity currents, 1. Motion of the head. Can J Earth Sci 3:523-546

Middleton GV, Southard J (1978) Mechanics of sediment movement. Binghamton, New.York.: Eastern Section, SEPM Short Course 3, 243 pp

Miller MC. McCave N. Komar P (1977) Threshold of sediment motion under unidirectional currents. Sedimentology 24:507-527

Milner HB (1962) Sedimentary petrography. Allen & Unwin, London, 715 pp

Munk W (1955) The circulation of the oceans. Readings from Scientific American: Oceanography. Freeman San Francisco, pp 64-69

Nikuradse J (1933) Strömungsgesetze in rauhen Rohren. Verein Deutsch Ing, Forschungsheft 361:1-22

Pettijohn F (1957) Sedimentary rocks (2nd ed). Harper, N.Y. 718 pp

Pickering K et.al.(1986) Deep water facies, processes and models. Earth Sci Rev 23: 75-174

Piper D et al.(1985) Sediment slides and turbidity currents on the Laurentian Fan: sidescar sonar investigations near the epicenter of the 1929 Grand Banks earthquake. Geology 13:538-541

Popper K (1981) Conjectures and refutations. Reprint. Routledge, London, 431 pp

Prandtl L, Oswatisch K, Wieghardt K (1969) Strömungslehre. Vieweg, Braunschweig 535 pp

Pratt JH (1859) On the influence of the ocean on the plumb-line in India. Philosophical Trans 149:779-796

Pray L, Murray R (eds)(1965) Dolomitization and limestone diagenesis. SEPM Special Publ 13, 180 pp

Rouse H, Ince S (1954) History of hydraulics. La houille blanche, Grenoble, 218 pp

Russell RD (1937) Mineral composition of Mississippi River sands. Bull Geol Soc Am 48:1306-1348

Schlüchter Ch(ed) (1979) Moraines and varves. Balkema, Rotterdam, 441 pp

Sclater J, Francheteau J (1970) The implications of terrestrial flow observations on current tectonic and geochemical models of the crust and Upper Mantle of the Earth. Geophys J R Astr Soc 20:509-542

Shackleton N, Kennett J (1975) Late Cenozoic oxygen and carbon isotopic changes at DSDP Site 284. Initial Reports of the Deep Sea Drilling Project 29:801-808, Washington, D.C.: U.S.Gov't Printing Office

Shepard F (1967) Submarine geology. 2nd ed, Harper NY, 558 pp

Shields A (1936) Anwendung der Aehnlichkeitsmechanik und der Turbulenzforschung auf die Geschiebebewegung. Mitteilungen d. Preuss. Versuchsanstalt f. Wasserbau u. Schiffbau, Berlin 26:1-26. The report was translated into English and issued as Report no 167 of the Keck Laboratory of Hydraulics, California Institute of Technology, Pasadena, California

Shreve R (1968) The Blackhawk landslide. Geol Soc Am, Spec Pap 108:1-47

Simons DB, Richardson EV, Nordin CF (1965) Sedimentary structures generated by flow in alluvial channels. SEPM Spec Publ 12:34-53

Smith GE (1980) Material-transfer involved in depositing the stratiform copper deposits of North Texas. In: Wolf KH (ed) Handbook of strata-bound and stratiform ore deposits 6:407-447

Stommel H (1958) The circulation of the abyss. In: Readings from the Scientific American, Oceanography. Freeman, San Francisco, pp 139-144

Sundborg Å (1956) The river Klarälven; A study of fluvial processes. Geograf Ann 38:125-316

Sverdrup HU, Johnson MW, Fleming RH (1942). The oceans. Prentice Hall, New York, 1087 pp

REFERENCES

Urey H, Lowenstam H, Epstein S, McKinney CR (1951) Measurement of paleotemperatures and temperatures of the Upper Cretaceous of England, Denmark, and the southeastern United States. Geol Soc Am Bull 62/4:399-416

Walker R (1973) Mopping up the turbidite mess. In: Ginsburg R Evolving concepts in sedimentology. Pettijohn Festschrift.pp 1-37

Zenger D, Dunham J, and Ethington R (eds) (1980) Concepts and models of dolomitization. SEPM Spec Publ 28, 320 pp

Zhao XF, Hsü KJ, Kelts K (1984) Varves and other laminated sediments of Zübo. Stuttgart, Schweizerbart, Contributions to Sedimentology 13: 161-176

Index